In this book, Adam Ferner frames challenges to my work which I take very seriously – not least with respect to the relation of being a person and being a human being. He seeks also to repair the acknowledged omission on my part to pursue the relation between the issues in logic and metaphysics I do discuss and profound issues in the philosophy of biology which I have neglected. I am surprised and gratified by the attention he has paid to my work.

David Wiggins, *New College, Oxford, UK*

There is a great deal of contemporary debate about the relation between metaphysics and the philosophy of biology, but it is surprisingly little remarked that no less a metaphysician than David Wiggins has been insisting on the importance of this relationship for half a century. In this book, Adam Ferner not only draws our attention to this fact, but goes on to explore the connection between Wiggins's ideas, especially on personal identity, and recent thinking in the philosophy of biology, in unprecedented depth. The book will be of great value not only to philosophers interested in personal identity and biological individuality, but also to many others interested in a fresh and important perspective on the work of Wiggins.

John Dupré, *University of Exeter, UK*

# Organisms and Personal Identity

Over his philosophical career, David Wiggins has produced a body of work that, though varied and wide-ranging, stands as a coherent and carefully integrated whole. In this book, Ferner examines Wiggins's conceptualist-realism, his sortal theory **D** and his Human Being Theory in order to assess how far these elements of his systematic metaphysics connect.

In addition to rectifying misinterpretations and analysing the relations between Wiggins's works, Ferner reveals the importance of the philosophy of biology to Wiggins's approach. This book elucidates the biological anti-reductionism present in Wiggins's work and highlights how this stance stands as a productive alternative to emergentism. With an analysis of Wiggins's construal of substances, specifically organisms, the book goes on to discuss how Wiggins brings together the concept of a person with the concept of a natural substance, or human being.

An extensive introduction to the work of David Wiggins, as well as a contribution to the dialogue between personal identity theorists and philosophers of biology, this book will appeal to students and scholars working in the areas of philosophy, biology and the history of Anglophone metaphysics.

**A.M. Ferner** is a research fellow at the SPH centre in Bordeaux and an Officer of the Royal Institute of Philosophy.

**History and Philosophy of Biology**
Series editor:
Rasmus Grønfeldt Winther | rgw@ucsc.edu | www.rgwinther.com

This series explores significant developments in the life sciences from historical and philosophical perspectives. Historical episodes include Aristotelian biology, Greek and Islamic biology and medicine, Renaissance biology, natural history, Darwinian evolution, Nineteenth-century physiology and cell theory, Twentieth-century genetics, ecology, and systematics, and the biological theories and practices of non-Western perspectives. Philosophical topics include individuality, reductionism and holism, fitness, levels of selection, mechanism and teleology, and the nature–nurture debates, as well as explanation, confirmation, inference, experiment, scientific practice, and models and theories vis-à-vis the biological sciences.

Authors are also invited to inquire into the 'and' of this series. How has, does, and will the history of biology impact philosophical understandings of life? How can philosophy help us analyse the historical contingency of, and structural constraints on, scientific knowledge about biological processes and systems? In probing the interweaving of history and philosophy of biology, scholarly investigation could usefully turn to values, power, and potential future uses and abuses of biological knowledge.

The scientific scope of the series includes evolutionary theory, environmental sciences, genomics, molecular biology, systems biology, biotechnology, biomedicine, race and ethnicity, and sex and gender. These areas of the biological sciences are not silos, and tracking their impact on other sciences such as psychology, economics, and sociology, and the behavioral and human sciences more generally, is also within the purview of this series.

**Rasmus Grønfeldt Winther** is Associate Professor of Philosophy at the University of California, Santa Cruz (UCSC), and Visiting Scholar of Philosophy at Stanford University (2015–2016). He works in the philosophy of science and philosophy of biology and has strong interests in metaphysics, epistemology, and political philosophy, in addition to cartography and GIS, cosmology and particle physics, psychological and cognitive science, and science in general. Recent publications include 'The Structure of Scientific Theories', *The Stanford Encyclopaedia of Philosophy* and 'Race and Biology', *The Routledge Companion to the Philosophy of Race*. His book with University of Chicago Press, *When Maps Become the World*, is forthcoming.

# Organisms and Personal Identity

Biological individuation and the work of David Wiggins

A.M. Ferner

 Routledge
Taylor & Francis Group

LONDON AND NEW YORK

First published 2016
by Routledge

2 Park Square, Milton Park, Abingdon, Oxfordshire OX14 4RN

52 Vanderbilt Avenue, New York, NY 10017

*Routledge is an imprint of the Taylor & Francis Group, an informa business*

First issued in paperback 2019

*British Library Cataloguing in Publication Data*
A catalogue record for this book is available from the British Library

*Library of Congress Cataloging in Publication Data*
A catalog record for this book has been requested

ISBN: 978-1-8489-3573-0 (hbk)
ISBN: 978-0-367-35861-7 (pbk)

Typeset in Times New Roman
by Wearset Ltd, Boldon, Tyne and Wear

For Esther

# Contents

# Acknowledgements

I have accrued a great many debts over the years it has taken me to write this book. Most prominent, of course, is the debt I owe to David Wiggins. His texts have proved a sure foundation for my own, and his comments have provoked and enriched my thoughts in exactly the right measure.

Thanks are due too to the members of the Birkbeck philosophy faculty. This book began life as a PhD thesis and I am grateful to those who assumed supervisory roles at points during my studies: Sarah Patterson, Robert Northcott, Ian Rumfitt, Keith Hossack and Susan James. Paul Snowdon deserves special mention both as a sympathetic extra-mural supervisor and as the BA tutor who first ignited my interest in Wiggins. Most of all, I have been fortunate that Jennifer Hornsby was the one to guide me through the latter stages of my PhD; she did so with characteristic insight and wit – and my work has been immeasurably improved by her suggestions. Comments from John Dupré and Rory Madden – my two examiners – have, alongside their other contributions to philosophy, improved the state of the arguments contained herein.

During the doctorate, I was helped as well by the humour and forbearance of my fellow students, in particular Robert Craven, Alex Douglas, Karl Egerton, Elianna Fetterolf, Charlotte Knowles, Tom Quinn, Christoph Schuringa and Neil Wilcox. Further afield, Nick Jones, Matteo Mossio, Gregory Radick, Ásta Sveinsdóttir and Charles T. Wolfe have helped me precisify my ideas about Wiggins and about organisms. Thanks are due to Anthony O'Hear and James Garvey, at The Royal Institute of Philosophy, for providing me with gainful employment by which I could fund my studies.

The process by which the thesis was transformed into a book was made much easier by the generous post-doctoral funding I received from Bordeaux-Montaigne University, and the year-long post I enjoyed there. The SPH department provided me with the requisite intellectual stimulation and I am grateful to my colleagues – Valéry Laurand, Pascal Duris, Steeves Demazeux, Céline Spector, Jauffrey Berthier, Adeline Barbin and Marion Bourbon – for their encouragement. I have Cédric Brun to thank, above all, for getting me to Bordeaux in the first place and for his stimulating conversations and continual support. I was also aided by Thomas Pradeu, who has helped create the conceptual space for works such as this one to exist, and with whom it has been a great pleasure to work. Thanks are due too to Sophie Gerber, and to Jean-François Moreau for the loan of his office at a crucial moment in the writing process.

On the editorial side of things, I issue my thanks to the series editor, Rasmus Grønfeldt Winther, for his insightful comments on the manuscript. A debt of gratitude is also owed to the two anonymous reviewers who assessed the draft and gave useful suggestions. Sophie Rudland helped with the initial preparation of the manuscript and Emily Briggs has effectively guided it through to completion.

Beyond the world of academia, debts were incurred on a larger scale to Flo, Luke, Viv and Mya who unwittingly took upon themselves the task of preserving my sanity: to a large extent they seem to have succeeded. I am grateful to my parents, Celia and Robin, for unwavering moral support. I would like to thank my sister, Harriet, for her guidance with (non-Aristotelian) psychology and my brother, Dave, for his advice some decade ago to pursue philosophy and for use of his PS3 (and to Medi for her tolerance of the same). Lastly, but most importantly, thanks are due to Esther, who has sustained me, and reassured and suffered me more than any person should reasonably be expected to.

# Introduction

In an autobiographical aside in *Identity, Truth and Value*, David Wiggins tells us that his earliest intellectual impulse was not towards philosophy, but to painting. And on more than one occasion while writing this book I have wondered – perhaps somewhat wistfully – whether his thoughts would have been better captured in that less restrictive medium. For whatever his written work is – beautifully crafted, subtle, ingenious – nobody would ever say it is easy to read.[1]

Wiggins himself laments the fact that his texts can be found obscure;[2] and a cursory glance at the numerous reviews of his books – *Sameness and Substance* and *Sameness and Substance Renewed* – suggests the view is widely held.[3] His prose is elegant, but it is a baroque elegance: dense and rich, with intricate and finely wrought digressions, and sentences that are, as Adrian Moore remarks, almost beyond parody in length.[4] There is an added – and not unimportant – difficulty for younger scholars approaching his work, following as it does a trajectory marked out more than forty years ago in *Identity and Spatio-Temporal Continuity*;[5] the trends in Anglophone philosophy have changed, and each new rendering of his theory is laden with references to older debates, while simultaneously incorporating responses to more recent ones.[6]

Yet the difficulties found in Wiggins's texts are also indicators of their strength. Thus we find the view, voiced by Peter Strawson[7] and echoed elsewhere,[8] that patience with Wiggins will be rewarded; the insights his work offers are both powerful and sustainable. Additionally, his forest of articles and books is valuable as *historical* markers: having been developed over so many decades they allow one to get a clear sense of the fluctuations of a particular moment in Anglophone philosophy. Every sentence carries marks of a myriad of past debates, in the form of footnotes, parenthetical comments and appendices.

Complex, confusing, rich and revealing – his work merits closer attention, and one aim of this book is to offer an introduction to it, albeit a partial one. Some aspects of my reading of Wiggins may seem surprising (the phenomenological rendering of his conceptualist-realism, for instance) while others will exhibit certain inadequacies; that my survey *is* partial may well worry those like David Bakhurst who emphasize the systemic nature of Wiggins's philosophy.[9] The focus here is on Wiggins's metaphysics, specifically his theory of individuation, his conceptualist-realism and his account of personal identity. His essentialism will be mentioned, but not analysed in depth.[10] His work in moral theory, semantics, aesthetics and ancient philosophy will be touched upon only briefly.[11] Because of space constraints some real and significant links between the various areas of his work will be neglected.

Yet, where some are neglected, other connections will be unearthed. The aim of the second half of this study is to trace the subterranean strands that link Wiggins's metaphysics to discussions in biology and the philosophy of biology. These links are equally real, and equally significant. It is no accident that, in the preface to *Identity and Spatio-Temporal Continuity*, Wiggins writes compellingly about the centrality of biology to metaphysical investigation:

> It gradually became evident to me in constructing this work that for the future of metaphysics no single part of the philosophy of science was in more urgent need of development than the philosophy of biology.[12]

It is a forceful pronouncement – one completely ignored by Wiggins's critics and commentators. This is an oversight which the present work attempts to address. It is argued below that a considerable amount of overlap exists between Wiggins's metaphysical themes and issues of biological individuality and anti-reductionism. How these biological issues bear on his 'Human Being Theory', and the 'personal identity debate' more generally, is examined in detail in the following chapters.

Before embarking on this project, however, some scene-setting is required; in the remainder of this introduction a brief overview will be given of the personal identity debate and the ambit of 'philosophy of biology' will be drawn, alongside a prospectus of the book as a whole and a short literature review. It should be noted that while, in some monographs, the chapters are written to be self-standing, this is not the case in the present work. The chapters are dependent on each other, as are the organs of a body, and I have tried my best to signpost the links between them in a way that I hope falls short of excessive.

## Personal identity

Questions about *what we are*, and *what it takes for us to persist*, are found in various forms, in multifarious philosophical traditions, from Buddhist perspectives on the self (or the lack thereof),[13] to conceptions of personhood in Yoruba thought.[14] On the whole, this work examines these questions as they appear in Anglophone philosophy. More specifically, the 'personal identity debate' is taken to refer to the body of discourse that has its historical roots in John Locke's discussion in the *Essay Concerning Human Understanding*.[15] At times, my analysis will move beyond this local setting (the discussion of Balinese conceptions of 'person' in Chapter 5 is an example), but in general, my discussion of 'personal identity' is organized in relation to that text. Artificial as this limit is, I take Wiggins's explicit reference to Locke as a good indicator of this point of departure and his attitudes towards these questions.

Locke's well-known treatment appears in Chapter 27 of Book II of his *Essay*: it is there that he claims that we – you and I – are fundamentally *persons*, where a person is 'a thinking intelligent being, that has reason and reflection, and can consider itself as itself, the same thinking thing, in different times and places'.[16] This definition lies at the core of his account of personal identity; it allows him to state what it takes for persons at earlier and later times to be identical and, consequently, to say what constitutes a person's persistence over time. For Locke, an individual found earlier is the same person as an individual found later if they are linked by *continued consciousness*, a psychological connection evidenced by experiential memory.

It is along these lines that Locke responds to the array of puzzle cases which have come to characterize the personal identity literature: stories of character transmigration (the 'prince and the cobbler' narrative),[17] of physical disassembly and reassembly (as in Christological accounts of bodily resurrection),[18] and – more prosaically but no less importantly – of memory-loss (exemplified in Locke's examinations of the 'drunkard').[19] Can persons survive such things? If not, why not? In recent years there have been some notable (and notably peculiar) additions to the canon – tales of teleportation[20] and fission[21] abound. Perhaps the most notorious of these additions is Sydney Shoemaker's 'brain transplantation' narrative,[22] in which Shoemaker describes the misfortunes of the sorry patient, Brown, who has his brain accidentally transferred into another patient's body. Asking whether or not Brown survives the transplantation, Shoemaker finds in Locke an affirmative answer. Neo-Lockeans[23] hold that Brown *does* survive, since his consciousness continues (so that the recipient of the brain donation can remember the past life of the donor, displays his personality, thoughts and so forth). This 'thought experiment' is the main focus of Chapter 9.

There are, of course, those who reject the Lockean analysis.[24] Neo-Lockeans or 'psychological theorists' are often contrasted with 'biological theorists' or 'animalists'.[25] Animalists such as Eric Olson deny that we are fundamentally persons. They hold that we are fundamentally 'human animals' and claim that our persistence conditions are those of the human animals that we are.[26] Whether or not we survive depends *not* on a psychological relation but on the continuation of *biological life*. Thus, for example, the animalist denies that the patient survives the brain transplantation.[27] Brain transplantation does not constitute the transferral of Brown into another body any more than transplants of his liver or kidney would.

There are those, however, who do not fit so easily into this binary between 'biological' and 'psychological' theorists – and David Wiggins is one. His considered view is that we are both fundamentally persons *and* fundamentally animals (or 'human beings'),[28] and that the terms 'person' and 'human being' are 'conceptually concordant'. Our understanding of one concept intimately involves our understanding of the other. In claiming this, his position can be read as 'psychological' in a much older sense of that word. Aristotle's notion of *psuche* or *psyche* is that of the form of the living human being (*empsuchon*), which encapsulates both its biological *and* cognitive nature. In many ways, Wiggins's account rehabilitates the insights of this Aristotelian picture. Exposition of his application of Aristotle to questions of personal identity is another central aim of this book.

## The philosophy of biology

While 'person' stands as a somewhat slippery term, nobody doubts that they have a ready grasp of what an animal is and where the spatio-temporal boundaries of such things lie.[29] Such, at least, is the consensus among those who discuss personal identity: animals are easy to track. This consensus, however, is not found beyond these confines. Turning to the philosophy of biology one finds these thoughts achieve a more controversial status. In this field, a wide array of conceptual puzzles within biological science are subjected to philosophical analysis.[30] Moreover, biological claims are brought to bear on traditional philosophical questions.[31] The concepts of biological *fitness* and *function* are the subject of considerable debate,[32] as is the status of so-called 'natural kinds'.[33] The focus below is particularly upon discussions of *biological individuality* and *metaphysical anti-reductionism*.

Biological individuality is the first, clearest point of contact between debates in 'personal identity' and the philosophy of biology. Both Wiggins and animalists of Olson's ilk rely on there being some suitably precise scientific method for picking out – 'individuating' – humans (the beings that they claim we are). Here, philosophers working within the field of 'biological individuality' are apt to find something troublesome.[34] While organisms like humans register clearly at the phenomenal level, the organismic divisions of the natural world are not obviously borne out at the more precise levels of biological theory. Phenomenal individuation, which seems to function relatively well with higher vertebrates, is ineffectual when it comes to individuating e.g. fungi and slime moulds.[35] Functional integration, which has been a popular model for determining organismal boundaries since Kant, has been criticized for its vagueness and for relying too heavily on our everyday assumptions about functional boundaries.[36] *Genetic* definitions are similarly confounded by colonial creatures and by clonal units.[37]

A variety of alternative accounts – phenomenal, epigenetic and immunological (and more) – are now on offer. Wiggins's project relies on there being a workable theory of organismic individuation by which the human organisms of everyday experience can be picked out and tracked.[38] The same applies to the 'animalists'. Whether or not there is such a theory is examined in Chapter 6.

One further point of contact between the philosophers of biology and the personal identity theorists is debate about the existence or 'reality' of *organisms*. Wiggins, like Aristotle, holds that organisms – humans not least – belong in the category of substance. Animalists agree.[39] Yet metaphysical (if not epistemic) reductionism is the majority view in the philosophy of biology, and organisms are often taken to be 'nothing more' than the fundamental particles that constitute them.[40] In what follows it will be proposed that the philosophers of biology will do well to precisify their discussion of metaphysical reductionism. Here is an instance in which philosophers of biology (if not biologists themselves) will benefit, as personal identity theorists surely do, from dialogue between these two spheres.

## Prospectus

The aim of the first three chapters is largely exegetical. In Chapter 1, I start by setting out Wiggins's overall approach to philosophical investigation. I situate him within a Strawsonian 'descriptive' tradition where metaphysical inquiry is guided by investigation of our pre-theoretical thoughts. His work is characterized by an emphasis on *elucidation*[41] (rather than reductive analysis of our everyday thinking) and by careful examination of the way our concepts develop *reciprocally*.[42] The descriptive approach is open to misgiving, especially in its association with *conceptualism*. Wiggins's response to these objections – realized in his open-minded 'conceptualist-realism' – is outlined. It will be argued that, while falling outside the mainstream in English-language philosophy, descriptivism is far from obviously false.

Once these methodological foundations are in place, discussion will turn in Chapter 2 to Wiggins's **D** theory, which encapsulates his interrelated claims about *identity* and *individuation*. His position is a 'sortalist' one; he claims that identity judgements can be reliably made only once it is specified what the items under investigation *are*. This is his sortal theory of *identity*. Being able to pick out and re-identify items similarly depends on picking them out as a *sort* of thing, with a specifiable *mode of being* or *principle of activity*. This is Wiggins's sortal theory of *individuation*. The exposition of **D** presented in Chapter 2 is developed in response to Paul Snowdon's reading. It corrects some of the problems with Snowdon's treatment by putting due emphasis on the reciprocity between our concept of identity and our everyday individuative practices.

Chapter 3 focuses on the *natural/artefactual* distinction as it figures in Wiggins's system. Some commentators – Massimiliano Carrara, Pieter Vermaas, Michael Losonsky and Lynne Rudder Baker – claim that Wiggins understands artefacts to have an impoverished ontological status in comparison to natural items. These interpretations are found to be wanting, resulting as they do from terminological confusions. An alternative reading is suggested, according to which artefacts are construed as metaphysically distinct from natural substances without necessarily being inferior. The notion of 'ontological dependence' is introduced here, for further elaboration in Chapters 8 and 9.

Chapter 4 expounds Wiggins's account of personal identity. Though interpreted by some (e.g. Olson) as 'neo-Lockean' and as 'animalist' by others (e.g. Peter Unger), Wiggins's work resists these readings and the first section of this chapter is focused on exactly how it straddles the 'psychological/biological' division. The thought that sits at the heart of his neo-Aristotelian picture is that the concepts *person* and *human being* are in some way, non-accidentally concordant. In the second section of this chapter, Wiggins's attempts to elucidate this conceptual consilience are drawn out. The connection is formed of three interwoven strands: the *Strawsonian argument*, the *semantic argument* and the *argument from interpretation*. While Wiggins has never aspired to lay down a transcendental proof for the conceptual concordance of *person* and *human being*, these elucidations show why the connection between them may be non-accidental and strong.

The aim of the fifth chapter is to test the strength of this conceptual connection – and it is suggested that a *genealogical* analysis of our notion of 'a person' undermines Wiggins's *semantic argument*. A genealogical sketch of that concept is offered (based on Marcel Mauss's essay 'A Category of the Human Mind'), and it is shown how our everyday use of the term relies on distinct and sometimes conflicting significations. It cannot give us insight into an unchangeable pre-theoretical structure. The *semantic argument* is undermined and the other two are weakened as a result. The conclusion reached, and revisited in Chapter 9, is that the connection between *person* and *human being* is not so steadfast as Wiggins supposes.

In the sixth chapter the claims of the first five chapters are situated in relation to issues in the philosophy of biology. Wiggins advises us to examine our (human) principle of activity through *biological inquiry*, and the intention here will be to organize the controversy by reference to biological individuation. There is a variety of models for picking out 'organisms' and Wiggins (as well as animalists) must decide which one best fits his purpose: *the physiological-functional account, the genetic account, the epigenetic account, the autonomy account* or *the immunological account*. Their relative strengths will be assessed and some provisional advice is offered (to be redeemed at the end of Chapter 9). Following this, some broader thoughts about Wiggins's verdant realism are offered. It is shown how he can respond to a criticism of 'genetic essentialism'.

After the interrogation of these biological models, another problem is raised in Chapter 7. All of the accounts presented in Chapter 6 construe the human being as constituted of, or composed of, or 'made up of', microscopic entities (like cells), which are themselves made up of even smaller entities (like quarks). The worry is that, in doing this, these models fail to match up with our pre-theoretical understanding. We do not see ourselves as composed of tiny living things – for the simple reason that they lie beyond our sense range. Attending to these miniscule entities encourages us to see divisions in the world that cut across the divisions of our everyday framework. Dramatically, this might motivate a *reductive* account (where organisms are 'nothing but' the tiny things that constitute them). The arguments for and against metaphysical reductionism are discussed in this chapter. The debate between reductionists and emergentists – like John Dupré – is laid out. An emergentist reading of Wiggins is offered, but in the end it is rejected.

In resisting reductionism, Wiggins need not pursue an emergentist line because, as is argued in Chapter 8, he has an alternative, generative form of neo-Aristotelian anti-reductionism at his disposal. Where reductionists and emergentists focus on *causal* dependence, Wiggins can focus on *ontological* dependence. Organisms are just as real as the stuff that makes them up because, in conceiving of them, we cannot help but understand them as *genuine unities*. They are *onto-logically prior* to their parts. This reading is presented in conjunction with an analysis of Aristotle's hylomorphism.

The final chapter of the book revolves around Shoemaker's brain transplanta-tion story and Wiggins's responses to this peculiar narrative. The chapter starts with a survey of his shifting attitudes towards it and describes how, despite numer-ous concerns, his assessment of the story remains unsatisfactory. Drawing on the arguments in the previous chapters it is shown why the narrative resists Wiggins's analysis. The contention is that Shoemaker's story is underpinned by a *mechanistic* logic and that it thus shifts one's metaphysical focus from *organisms* – natural sub-stances that are prior to their parts – to 'living something-or-others' – entities whose parts can be conceived of as separable from the whole. The diagnosis of Shoemaker's 'thought experiment' as mechanistic is supported by reference to Ian Hacking's analysis of transplantation and by a speculative historical study offered in the Appendix. These thoughts are further developed by reference to the discus-sion in Chapter 3 about the metaphysical character of *artefacts*. The notion of a 'living something-or-other' is precisified and the brain transplantation story is seen to describe the adventures of a *biological artefact* (not an organism). In conclud-ing, the implications for Wiggins's Human Being Theory are discussed and the cri-tique offered in Chapter 5 is reiterated and reinforced.

No tract on organisms would be complete without an Appendix. Mine sets out to supplement the arguments in Chapter 9 about the connections between the neo-Lockean position and biological mechanism. It is argued that, while the con-nection may not be necessary, it is strong and non-accidental; Locke's account of personal identity develops as a direct result of his endorsement of Boylean corpuscularianism. Shoemaker – who explicitly draws on the relevant passages in *The Essay Concerning Human Understanding* – inherits this conceptual alli-ance along with the claims about continuity of consciousness.

## Overview of texts

Some remarks on Wiggins's texts and the secondary literature should be entered here. The primary source for this research is *Sameness and Substance Renewed* (2001) (hereafter '*S&SR*').[43] This book – which contains his metaphysical accounts of substance, identity and personhood – is a modified and expanded form of *Sameness and Substance* (1980) ('*S&S*'),[44] which itself is a (consider-able) development of *Identity and Spatio-Temporal Continuity* (1967) ('*ISTC*').[45] In *S&SR* Wiggins aimed to respond to the issues raised by the first two books (and intervening articles) – the result, as noted, is a complex and often digressive work.[46] It has been succeeded by a spate of papers, including a helpful series of responses to a forum in *Philosophy and Phenomenological Research* in 2005, and a fiery discussion with Shoemaker in *The Monist* in 2004.[47] Also of interest is the *festschrift* compiled for Wiggins's 60th birthday (and to mark his acces-sion to the Wykeham chair of Logic): *Identity, Truth and Value* (1996).[48] Edited by Sabina Lovibond and S.G. Williams, it includes a series of papers on

Wiggins's contributions and his replies and his 'personal-cum-academic' memoir. More recently and perhaps most usefully, Wiggins's 2012 Mark Sack's lecture – 'Identity, Individuation and Substance' – has been published in the *European Journal of Philosophy*[49] – in this, he recognizes the need to restate his metaphysical position (for a new generation) and goes some way towards achieving this. I have also been lucky enough to see versions of his latest collection, *Twelve Essays*, forthcoming from Oxford University Press, which has proved immensely helpful. While I will be focusing primarily on *S&SR*, my reading is informed no less by *S&S*, the clearer of the two texts, and the later papers, which iron out some inconsistencies.

The reading presented herein is greatly dependent on three other texts, apart from those by Wiggins. The first is Aristotle's *Categories* (and, to lesser degrees, his *Metaphysics*, *Physics* and *De Anima*). Throughout his academic career, Wiggins has explicitly stated the centrality of Aristotelian thought to his metaphysical and ethical positions, and an understanding of Aristotle is crucial to the interpretation essayed in Chapter 8. The second text is Peter Strawson's *Individuals: An Essay in Descriptive Metaphysics*.[50] Strawson is a silent partner in Wiggins's project; he is mentioned infrequently – yet the influence of his particular 'descriptive' approach to metaphysics pervades Wiggins's work. The intellectual debt owed here is examined in depth in Chapter 1. There is a third text that should be included alongside these – Gottfried Leibniz's *New Essays on Human Understanding*.[51] Wiggins takes over Leibniz's differentiation of *clear* (ordinary, usable, but provisional) ideas from *distinct* ideas (ideas that are philosophically or scientifically analysed). Moreover his account of individuation is visibly indebted to Leibniz's conception of the activity of a substance.[52]

As regards to secondary literature, I have been helped by Paul Snowdon's papers: 'Persons and Personal Identity',[53] and 'On the Sortal Dependency of Individuation Thesis'.[54] Jonathan Lowe's discussion of Wiggins's 'conceptualist-realism' in the *Philosophy and Phenomenological Research* forum has also been particularly instructive, as has Bakhurst's contribution to the same.[55] S.G. Williams' entry in the *Continuum Encyclopedia of British Philosophy* is noteworthy for emphasizing, appropriately, the modesty of Wiggins's philosophical approach; how he aims, not for conceptual analysis, but for the *elucidation* of central, human concepts.[56] It is unfortunate that despite the numerous reviews and articles about Wiggins's work there is still no sustained exposition of his theories other than his own – and this is another oversight that the present work hopes to address.

## Notes

1 See Wiggins 1996: 222.
2 Wiggins 2012: 1.
3 '[One] has to struggle with convoluted sentences and a certain elusiveness of argumentative structure' (Lowe 2003: 816), '[It is] set down in a rather haphazard organization of main text, subsidiary text, footnotes, and further notes' (Baldwin 1982: 270),

> *Sameness and Substance* is a difficult book; Wiggins's avowed effort to be understood will be lost on those who are not prepared to follow the intricacy of his response to criticism and to look for the relevance of papers written since 1967 and listed in the preface.
>
> (Cartwright 1982: 597)

See also Noonan 1981.

4  Moore 1996: 165.

5  See Lowe 2003: 816 (and Cartwright 1982: 597) for elaboration.

6  Bakhurst (2005) offers a short review of some of the changes.

7  Strawson 1981: 603.

8  E.g. Noonan 1981: 261, Cartwright 1982: 597, Lowe 2003: 816, Snowdon 2009: 254.

9  Bakhurst 2005. See also Lovibond and Williams 1996: *Preface*.

10  For a closer study of this I recommend Nick Jones's 'Individuation to Essentialism' (*draft*).

11  And, sadly, a language barrier prevents proper engagement with those works he has produced in French.

12  Wiggins 1967: vii.

13  See Ganeri 2012 for an interesting attempt to nurture dialogue between the questions as they feature in the Buddhist and Anglophone spheres.

14  E.g. Adeofe 2004.

15  Locke 1690/1975.

16  Ibid.: II, xxvii, §9.

17  Ibid.: II, xxvii, §15.

18  This is discussed in depth in Chapter 4.

19  Locke 1690/1975: II, xxvii, §21 and §23.

20  E.g. Parfit 1984.

21  E.g. Shoemaker 2004a.

22  Shoemaker 1963: 23–24.

23  There have been considerable refinements to the Lockean position in recent years. Neo-Lockean revisions – such as Shoemaker's (e.g. 1963, 2004), and Parfit's (e.g. 1971, 1984) – attempt to re-describe the psychological relation without putting Locke's (problematic) emphasis upon memory-links. For discussions of the problems in Locke, see, for example, Reid 1785, Essay III, especially Chapters 3 and 4 (for a more recent survey, see Noonan 1989). Neo-Lockeans also typically endorse some form of materialism according to which consciousness is inextricably tied to some material substrate, the human brain.

24  One reason is its failure to match our everyday intuitions about the foetus which we believe we once were.

25  A distinction found in e.g. Noonan 1998 and Olson 1997.

26  For a classic statement of this position see Olson 1997. Besides Olson, the term 'Animalism' has been used to describe the view of W.R. Carter (1989), Michael Ayers (1991), Trenton Merricks (2001), Paul Snowdon (1990) and Rory Madden (*draft*).

27  A distinction is drawn – and should be mentioned here – between *brain transplants* and *cerebrum transplants*. As Olson points out, the cerebrum is 'the organ that is most directly responsible for higher mental capacities' (Olson 1997: 9). Thus the transposition of the cerebrum would cover the transfer of psychological features like experiential memory. The cerebrum, however, does not include the *brain stem*; 'the organ that is chiefly responsible for directing your life sustaining functions' (1997: 140) (e.g. respiratory, digestive, metabolic processes). For the sake of the arguments to follow, the focus here is on brain transplantation, where both cerebrum and brain stem are supposed to be transposed (for reasons which will become evident in Chapters 3 and 4).

28  Wiggins 2001: 193. The difference between 'human being' and 'human animal' is set out in Chapter 4.

29  Wiggins 2001: 193.
30  Philosophers of biology often say that their field is a relatively young sub-discipline of philosophy. In their influential introductions to the subject both Michael Ruse and David Hull claim that philosophical analyses of biological concepts are scarce in the early half of the twentieth century (Ruse 1973 and Hull 1974), and – as Alexander Rosenberg notes – any thoughts about biology by philosophers were included primarily as 'an afterthought to discussions of physics' (Rosenberg 1985: 6–7). The quotation above, from Wiggins 1967, attests to a growing demand for philosophical development in that area – but what Ruse, Hull and Rosenberg's surveys occlude is the important, though side-lined, research that *was* produced prior to 1960 (for an overview of these issues see Nicholson and Gawne 2013). One figure that escapes their articulation of the subject is J.H. Woodger. In Chapters 5 and 6 Woodger's influence on Wiggins will be emphasized. On Woodger, see Nicholson and Gawne 2013.
31  See Griffiths 2011 for a good overview.
32  See Hull and Ruse 1998 (1ff) for an introduction to these issues.
33  See, for example, Dupré 1993.
34  E.g. Wilson 1999, Pradeu 2010, Clarke 2013.
35  See Hull 1992.
36  Pradeu 2010.
37  Sterelny and Griffiths 1999: 71.
38  This point is made in Wilson 1999.
39  Following van Inwagen 1990 (see Olson 2007).
40  See, for example, Hull 1992.
41  This aspect of his work is drawn out in Williams 2006.
42  Wiggins 1996:

> Indeed reciprocity – or two-way flow, as I used to call it in internal dialogue with myself – was part of a more general thought.... It seemed integral to the proper understanding of Meno's dilemma.... It seemed indispensable to the proper understanding of what we achieve when we come to know what a thing is. It was integral, in a way still too little-heeded or thought through by the philosophy of science, to our understanding of thing-kind words like 'horse' or 'human being'. And not only that. It was a further generalization of the reciprocity point that helped to make it possible to contemplate new possibilities in connection with questions of value.
>
> (228)

43  Wiggins 2001.
44  Wiggins 1980.
45  Wiggins 1967.
46  Wiggins 2001: ix.
47  See Wiggins 2004a and 2004b, and Shoemaker 2004a and 2004b.
48  Lovibond and Williams 1996.
49  Wiggins 2012.
50  Strawson 1959.
51  Leibniz 1765/1981.
52  We might also mention Frege as a fourth source of inspiration, not least his account of sense and reference and the clarity that has flowed from Frege's work into twentieth century logic.
53  Snowdon 1996.
54  Snowdon 2009.
55  Lowe 2005, Bakhurst 2005. Stephen Yablo's review of *S&SR* for the *TLS* (2003) is also commendable for its clarity and brevity.
56  Williams 2006.

# Bibliography

Adeofe, L. (2004) 'Personal Identity in African Metaphysics', in L. Brown (ed.) *African Philosophy: New and Traditional Perspectives* (Oxford: Oxford University Press).

Aristotle. (1936) *Physics,* W.D. Ross (ed. and trans.) (Oxford: Clarendon Press).

Aristotle. (1938) *Categories,* H.P. Cooke and H. Tredennick (ed. and trans.) (Cambridge, MA: Harvard University Press).

Aristotle. (1961) *De Anima,* W.D. Ross (ed. and trans.) (Oxford: Clarendon Press).

Aristotle. (1994) *Metaphysics* (Books Z and H), D. Bostock (ed.) (Oxford: Clarendon Press).

Ayers, M. (1991) *Locke,* vol. 2 (London: Routledge).

Bakhurst, D. (2005) 'Wiggins on Persons and Human Nature', *Philosophy and Phenomenological Research* 71(2): 462–469.

Baldwin, T. (1982) 'Review of *Sameness and Substance*', *Philosophy* 57(220): 269–272.

Carter, W.R. (1989) 'How to Change your Mind', *Canadian Journal of Philosophy* 91: 1–14.

Cartwright, H.M. (1982) 'Review of Sameness and Substance', *The Philosophical Review* 91(4): 597–603.

Clarke, E. (2013) 'The Multiple Realizability of Biological Individuals', *The Journal of Philosophy*, CX (8), 413–435.

Clarke, E. (*forthcoming*) 'The Multiple Realizability of Biological Individuals', *Journal of Philosophy*.

Dupré, J. (1993) *The Disorder of Things: Metaphysical Foundations of the Disunity of Science* (Cambridge, MA: Harvard University Press).

Ganeri, J. (2012) *The Self: Naturalism, Consciousness and the First-Person Stance* (Oxford: Oxford University Press).

Griffiths, P. (2011) 'Philosophy of Biology', in Edward N. Zalta (ed.) *The Stanford Encyclopedia of Philosophy* (Summer Edition), available online at: http://plato.stanford.edu/archives/sum2011/entries/biology-philosophy/

Hull, D. (1974) *Philosophy of Biological Sciences* (Englewood Cliffs, NJ: Prentice-Hall).

Hull, D.L. (1992) 'Individual', in E. Fox Keller and E. Lloyd (eds) *Keywords in Evolutionary Biology* (Cambridge, MA: Harvard University Press): 180–187.

Hull, D.L. and Ruse, M. (1998) (eds) *The Philosophy of Biology* (Oxford: Oxford University Press).

Jones, N. (*draft*) 'From Individuation to Essentialism'.

Leibniz, G. (1765/1981) *New Essays Concerning Human Understanding,* P. Remnant and J. Bennett (trans.) (Cambridge: Cambridge University Press).

Locke, J. (1690/1975) *An Essay Concerning Human Understanding,* P.H. Nidditch (ed.) (Oxford: Oxford University Press).

Lovibond, S. and Williams, S. (1996) *Essays for David Wiggins: Identity, Truth and Value* (Oxford: Blackwell Publishing).

Lowe, E.J. (2003) 'Review of *Sameness and Substance Renewed*', *Mind*, New Series, 112(October): 448.

Lowe, E.J. (2005) 'Is Conceptualist Realism a Stable Position?', *Philosophy and Phenomenological Research* LXXI(2): 456–461.

Madden, R. (*draft*) 'The Persistence of Animate Organisms'.

Merricks, T. (2001) 'How to Live Forever without Saving your Soul: Physicalism and Immortality', in K. Corcoran (ed.) *Soul, Body and Survival* (Ithaca: Cornell University Press).

Moore, A.W. (1996) 'On There Being Nothing Else to Think, or Want, or Do', in S. Lovibond and S. Williams (eds) *Essays for David Wiggins: Identity, Truth and Value* (Oxford: Blackwell Publishing).

Nicholson, D. and Gawne, R. (2013) 'Rethinking Woodger's Legacy in the Philosophy of Biology', *Journal of the History of Biology* 47: 243–292 (Springer).

Noonan, H. (1981) 'Review of *Sameness and Substance*', *The Philosophical Quarterly*, 31(124): 260–268.

Noonan, H. (1989) *Personal Identity* (London: Routledge).

Noonan, H. (1998) 'Animalism versus Lockeanism: A Current Controversy', *The Philosophical Quarterly* 48: 302–318.

Olson, E. (1997) *The Human Animal: Personal Identity Without Psychology* (Oxford: Oxford University Press).

Olson, E. (2007) *What Are We? A Study in Personal Ontology* (Oxford: Oxford University Press).

Parfit, D. (1971) 'Personal Identity', *Philosophical Review* 80: 3–27.

Parfit, D. (1984) *Reasons and Persons* (Oxford: Oxford University Press).

Pradeu, T. (2010) 'What is an Organism? An Immunological Answer', in P. Huneman and C.T. Wolfe (eds) *History and Philosophy of the Life Sciences*, special issue on *The Concept of Organism: Historical, Philosophical, Scientific Perspectives. History and Philosophy of the Life Sciences* 32(2–3).

Reid, T. (1785) *Essays on the Intellectual Powers of Man* (Edinburgh: John Bell).

Rosenberg, A. (1985) *The Structure of Biological Science* (Cambridge: Cambridge University Press).

Ruse, M. (1973) *The Philosophy of Biology* (London: Hutchinson & Co.).

Shoemaker, S. (1963) *Self-Knowledge and Self-Identity* (Ithaca: Cornell University Press).

Shoemaker, S. (2004a) 'Brown-Brownson Revisited', *The Monist* 87(4): 573–593.

Shoemaker, S. (2004b) 'Reply to Wiggins', *The Monist* 87(4): 610–613.

Snowdon, P. (1990) 'Persons, Animals and Ourselves', in C. Gill (ed.) *The Person and the Human Mind* (Oxford: Clarendon Press).

Snowdon, P. (1996) 'Persons and Personal Identity', in S. Lovibond and S. Williams (eds) *Essays for David Wiggins: Identity, Truth and Value* (Oxford, Blackwell Publishing).

Snowdon, P. (2009) 'On the Sortal Dependency of Individuation Thesis', in H. Dyke (ed.) *From Truth to Reality: New Essays in Logic and Metaphysics* (London: Routledge).

Sterelny, K. and Griffiths, P. (1999) *Sex and Death: An Introduction to Philosophy of Biology* (Chicago: University of Chicago Press).

Strawson, P.F. (1959) *Individuals: An Essay in Descriptive Metaphysics* (London: Methuen).

Strawson, P.F. (1981) 'Review of *Sameness and Substance*', *Mind*, New Series, 90(360): 603–607.

van Inwagen, P. (1990) *Material Beings* (Ithaca: Cornell University Press).

Wiggins, D. (1967) *Identity and Spatio-Temporal Continuity* (Oxford: Blackwell).

Wiggins, D. (1976) 'Locke, Butler and the Stream of Consciousness: And Men as a Natural Kind', *Philosophy* 51: 131–158.

Wiggins, D. (1980) *Sameness and Substance* (Cambridge, MA: Harvard University Press).

Wiggins, D. (1996) 'Replies', in S. Lovibond and S. Williams (eds) *Essays for David Wiggins: Identity, Truth and Value* (Oxford: Blackwell Publishing).

Wiggins, D. (2001) *Sameness and Substance Renewed* (Cambridge: Cambridge University Press).

Wiggins, D. (2004a) 'Reply to Shoemaker', *The Monist* 87(4): 594–609.

Wiggins, D. (2004b) 'Reply to Shoemaker's Reply', *The Monist* 87(4): 614–615.

Wiggins, D. (2005a) 'Précis of "Sameness and Substance Renewed"', *Philosophy and Phenomenological Research* 17(2) 442–448.

Wiggins, D. (2005b) 'Reply to Bakhurst', *Philosophy and Phenomenological Research* 17(2): 470–476.

Wiggins, D. (2012) 'Identity, Individuation and Substance', *European Journal of Philosophy* 20(1): 1–25.

Wiggins, D. (2016) *Twelve Essays* (Oxford: Oxford University Press).

Williams, S.G. (2006) 'David Wiggins', in N. Goulder, A.C Grayling and A. Pyle (eds) *Continuum Encyclopedia of British Philosophy* (London: Thoemmes Continuum).

Wilson, J. (1999) *Biological Individuality* (Cambridge: Cambridge University Press).

Yablo, S. (2003) 'Tables Shmables: Review of David Wiggins, *Sameness and Substance Renewed*', in the *Times Literary Supplement* (July).

# 1    An intellectual microcosm

## 1 Pre-history

Theories and counter-theories flourish in specific intellectual climates, and to study them effectively it is necessary to have a sense of the familiar or unfamiliar environment of which they are a part. To this end, the first section of this chapter describes the cultural and intellectual microcosm in which David Wiggins's ideas about individuation and organisms took root – specifically in P.F. Strawson's peculiarly fertile *descriptive metaphysics*.[1] Historical scholarship is not the primary aim of this book – as will be clear from the gestural nature of the account below – but I hope the purpose will be served.

In the second section, we will get a clearer idea of the distinctive features of Strawsonian descriptivism: its focus on linguistic analysis, its emphasis on ontological priority, and its systematicity. And in the third section, it will be shown how Wiggins inherits certain conceptual defects from his intellectual forebears, and some tentative suggestions for correcting them will be offered. Further concerns will be raised in the fourth section, and these will spiral into the chapters that follow. Many of the controversial aspects of this book arise from Wiggins's particular attitude to metaphysical analysis and it is for this reason that we will turn first to the pre-history, to the trends that gave rise to his distinctive approach to philosophy.

\* \* \*

Spontaneous generation exists in neither nature nor history, so any starting point to this story will inevitably be somewhat arbitrary. With that in mind, I begin some years before Peter Strawson's appearance on the scene, with Logical Positivism and the loose association of thinkers known as the Vienna Circle. Active in 1920s Austria, this philosophical collective – which included Rudolf Carnap, Otto Neurath and Moritz Schlick – is notable for positioning the empirical sciences at the centre of its philosophical practice. Their primary epistemological claim, broadly construed, was that one should honour only those facts that are logically or scientifically verifiable. Systems of knowledge, they held, should only discuss things that can be accounted as either true or false. Their view resulted in an important antipathy towards traditional metaphysical questions, questions about the nature of Being or the existence of God or other abstract entities.

In their manifesto, 'The Scientific World Conception' (1929), the Logical Positivists opposed the metaphysical discussions of Henri Bergson, F.H. Bradley and Martin Heidegger. Metaphysical claims cannot be verified by appealing to science, they said. Nor yet are such statements analytic – they are not true in virtue of the meanings of their constituent terms. And since they cannot be verified, the Positivists saw them as falling outside their intellectual programme. In their manifesto and works such as Carnap's 'Pseudo-problems in Philosophy' (1928), they decried metaphysical utterances as nonsensical pseudo-propositions. They are, they held, more akin to poetry; they seem impressive, of course, full of sound and fury, but ultimately they signify nothing.

Nor was this anti-metaphysical attitude the sole preserve of the Vienna Circle. It was A.J. Ayer's *Language, Truth and Logic* (1936) that imported this brand of Positivism into the Anglophone mainstream and spread these suspicions about nonsensical transcendental musings. Thus, commenting on the state of English-Language philosophy in the pre-war years, Gilbert Ryle writes:

> In the 1930s the Vienna Circle made a big impact on my generation and the next generation of philosophers. Most of us took fairly untragically its demolition of Metaphysics. After all we never met anyone engaged in committing any metaphysics; our copies of *Appearance and Reality* were dusty; and most of us had never seen a copy of *Sein und Zeit*.[2]

The contrast here, with the current rude health of Analytic metaphysics, is striking. Even those with only a cursory familiarity with contemporary academic philosophy will know that discussions of ontology and existence questions are flourishing (so much so that sub-disciplines like the ludically named 'metametaphysics'[3] have emerged to discuss its methodology and its successes). Metaphysics, the discussion of reality, the assessment of existence questions, has never been in a better state. It seems then, that news of its 'demolition' was greatly exaggerated. Or if it was demolished, the building has been rebuilt, reinforced, its halls re-inhabited.

Strangely enough, narratives of mid-twentieth century Analytic philosophy cite the rise of 'conceptual analysis' as a contributing factor in the return to metaphysical inquiry.[4] Conceptual analysis was the central method of the movement known as 'ordinary language philosophy'. This took hold in Oxford in the 1940s in the wake of work by G.E. Moore and Ludwig Wittgenstein. Its aim was to solve philosophical problems by reflecting on language and our everyday concepts. In the present history it stands as a significant precursor of Strawsonian-cum-Wigginsian descriptive analysis. What is strange is that while it eventually begat descriptive metaphysics, the 'ordinary language' school was at first powerfully allied with Logical Positivism. Ordinary language philosophy shared in the no-nonsense approach to philosophizing. They aimed to root out nonsense, and nonsensical questions, by clarifying features of our linguistic and conceptual systems. Ryle's work exemplifies this trend. In his book, *The Concept of Mind* (1949), he argued that the Cartesian dogma of the 'ghost in the machine' (according to which an immaterial thinking substance controls the body) is the consequence of a 'category-mistake'; conceptual categories have been confused – mental descriptions have been connected up with the language of physical events in a nonsensical fashion, producing the multifold problems of dualism, problems which are all the more dogged for being intangible. J.L. Austin is another representative of the 'ordinary language' attitude. Like Ryle, he thought careful attention to our everyday speech gives insight into, and hopefully resolves, traditional problems of philosophy (in particular, in *Sense and Sensibilia* (1962) he takes discussions of 'sense data' as his target).[5]

It is in the work of Peter Frederick Strawson that this current of conceptual analysis breaks against the anti-metaphysical waves flowing from Vienna; it is with Strawson that we find a determinate *return* to metaphysical questions in Anglophone philosophy. And it is P.F. Strawson who was perhaps most central in loaming the earth in which Wiggins was to later lay his theories of substance and individuation.[6] Like Ryle and Austin, Strawson saw the analysis of language to be central to philosophical practice. But his work differs from theirs in its scope and generality. Among the concepts it investigates some are taken to be basic, irreducible and general: concepts of *space* and *time*, of *material object*, of *property* and of *causation*.[7] Strawson aspired, as Peter Hacker puts it, 'to the degree of generality characteristic of the ontological and metaphysical pronouncements of the great system-builders of the past'.[8] His aim, realized in his 1959 monograph, *Individuals: An Essay in Descriptive Metaphysics*,[9] was to use linguistic analysis, not simply to identify conceptual inconsistencies, but as a method for divining the warp and weft of reality. He used conceptual analysis to pursue metaphysical inquiry.

Here, Strawson picks up a thread lain by one of the most celebrated of system-thinkers, Immanuel Kant, and reworks it into the fabric of post-war Anglophone philosophy. This, indeed, was one of his most influential, most lasting, innovations. Until Strawson's *Individuals* and *The Bounds of Sense*,[10] Anglophone Analytic interest in Kant had been largely historical in nature.[11] Strawson's re-working re-introduced him to the mainstream in a way that was surprisingly accessible to a metaphysically suspicious community. Kant is, perhaps, one of the last figures one would expect to find in an intellectual history of David Wiggins. All the same, via Strawson, Kant's thought can be seen to sustain the work found in *Sameness and Substance* and its sister texts. One of the central Kantian themes that Strawson took as his subject, and which in turn informs Wiggins's position, is the apparent distance between the human mind and reality. Kant denied that we have direct access to reality, to things in themselves. The *noumenal* realm, as he had it, lies forever beyond our reach. Thus, Kantian metaphysics is not focused on investigating raw, pre-conceptual entities or pursuing ontological searches for metaphysical essences. The world of forms does not submit to human study. What Kant aimed at investigating was *the structure of experience*. What we needed to understand, he thought, were the necessary preconditions for the possibility of *experiencing* reality.

> There is a difference between experiences and their objects, and the content of experience is contingent. But, according to Kant, there are also necessary or structural features of experience, and these determine the necessary or essential features of the *objects* of experience.[12]

This quotation, from Hans-Johann Glock's discussion of Strawsonian Kantianism (or Kantian Strawsonianism), points to the notorious idealist twist found in Kantian thought. As discussed below, 'idealism' is one of the knee-jerk criticisms levelled at the descriptivist philosopher, and one to which Wiggins has, at times, felt encouraged to respond. Strawson, following Kant, took the structural features of our experience to correspond to structural features of reality – and in doing so seemed to suppose that the external world is somehow *dependent* on epistemic conditions.[13] In the sections that follow, Wiggins's brand of descriptivism will be defended against this charge. The present aim is only to gesture toward his intellectual ancestry, to position Kant as one of his philosophical forebears and to give a sense, however vague, of the conceptual evolution of these ideas.[14]

Mapping these lines of descent also reveals links to not-so-distant philosophical cousins. The Phenomenological tradition – realized in the texts of (among others) Husserl, Hegel, Heidegger, Merleau-Ponty and Sartre – is another offshoot of these Kantian thoughts. And scholars of these figures will find unexpected echoes and reflections in Wiggins's work. It is indicative of the self-isolating tendencies of Analytic philosophy that the links between descriptivism and the Phenomenological tradition are not more widely discussed.[15] This is a connection that I hope will become clearer in the pages below.[16]

Strawson thought that the structure of our experience gave us insight into the structure of *reality* – but following the linguistic turn of the ordinary language philosophers he saw language as a means by which to get a sense of this experiential structure. For Strawson, like Kant, one could pursue metaphysical inquiry by examining the necessary features of experience. While lacking direct access to the things in themselves, we can come to understand them better by better understanding how we experience them. We experience physical objects as located in space and time, as undergoing change, as standing as nodes in causal networks. But sense has its bounds. We cannot understand material things outside these limits. Our ability to navigate the world in the wonderful way that we do is premised on this structure. The aim of the descriptivist metaphysician has to be to find and describe these necessary, basic structural features.

'The structure of experience' – it is awkward to reflect on something like this, which fills the reflection itself. It is like trying to see the back of one's head in a mirror. By contrast, consider the reassuringly prosaic analogy of *filing*. There are in life multifarious forms of administrative filing. In some cases it will be necessary to put cases in drawers, in others, papers in boxes, in others still, paper – glory abounds – in files, in boxes in drawers, on shelves and so on, right up to the secretary-in-chief's private filing-cabinet. The documents and reports and memoranda are all ordered according to some *system*. Each piece of paper is studied and, meeting a criterion, it is put in a file. Each file is itself studied and then placed in a box. And onwards. These systems can be used to transform the heaving morass of papers on the academic's desk into something that is comprehensible (so we are told). In the same way, we order the raw data of experience. We *categorise* to understand; we are constantly, and often unconsciously, squaring away the world: dog, cat, object, property, time, space.... For Kant and Strawson, we are born administrators. Our method for filing away the elements of reality is the subject of their analysis.

*    *    *

Peter Frederick Strawson is central to the post-war renovation of Anglophone metaphysics. He created a space for metaphysical analysis, despite prevailing Positivist trends. And he did this in a way that appealed to the ordinary language philosophy emerging from Oxford. He belongs in the pre-history of Wiggins's work on substance and personal identity.

Yet, for the wary reader a worry will remain. Strawson's metaphysics is a study of human understanding. It is about the boundaries of knowledge – and this sounds suspiciously epistemological in tone. Is that why it was so agreeable to the post-Positivist Anglophones? To what extent is descriptive metaphysics *really* metaphysics? How viable is it as a method for gaining a surer grasp on reality? In the following sections, the Strawsonian method is articulated in greater detail and the objections against it are considered.

## 2  Descriptive metaphysics

Metaphysics has been often revisionary, and less often descriptive. Descriptive metaphysics is content to describe the actual structure of our thought about the world, revisionary metaphysics is concerned to produce a better structure.... [Descriptive metaphysics] needs no justification at all beyond that of inquiry in general. Revisionary metaphysics is at the service of descriptive metaphysics. Perhaps no actual metaphysician has ever been, both in intention and effect, wholly the one thing or the other. But we can distinguish broadly: Descartes, Leibniz, Berkeley are revisionary, Aristotle and Kant descriptive.[17]

So starts Strawson's seminal text, *Individuals: An Essay in Descriptive Metaphysics*. The descriptivist – who works in the shadow of Aristotle and Kant – examines reality by looking to the way we *experience* it. Latent in this structure is our 'conceptual scheme',[18] reflecting (it is supposed) central features of the make-up of the world. The revisionary metaphysician, by contrast, is not content to study the categories of experience which she suspects to be merely local filing-systems rooted in a limited human perspective. Rather, she encourages us to reach beyond the bounds of the human mind, towards a more 'objective' reality. The world might *appear* to us to be a collection of material continuants (chairs, tables, cats, dogs ...) but this is an accident of our biology (if not our culture); for the revisionists, physical sciences show us how to improve our world-view – medium-sized objects are in fact nothing more than bundles of reactions connected by physical laws. A cat, far from being a discrete individual, is a heterogenous composite of numerous genetically diverse systems. So she may say. To echo a distinction essayed by the physicist Ernst Mach, there is a 'critically purified' picture of the world offered to us by science, and there is the 'vulgar' one of everyday business, a practical if unrefined approach which allows us to get by in day-to-day business.[19] The revisionary metaphysicians – 'process' philosophers like Whitehead, and 'four-dimensionalists' like David Lewis (as well as most broadly Quinean metaphysicians) – aim to study what reality is *really* like. The descriptivist aims to do the same, but does so by dealing with the 'vulgar'.[20]

We will return, in due course, to revisionary metaphysics, the criticisms it offers and its own somewhat dubious prospects. For now, however, let us remain with Strawson and consider how his descriptivism is taken up in Wiggins's work. The following sections position Wiggins within the descriptivist tradition and simultaneously assess how he measures up to the various criticisms of Strawson's form of descriptivism. To this end we can first examine, if briefly, Strawson's analysis of *reference*, which stands as a canonical example of the method.

*Reference* is the grammatical procedure that transfuses ordinary language whereby one object – a person, for example – designates, or picks out, or identifies, another, be it the page you are reading, the breath that sustains you or the colour red. And since his paper 'On Referring' (1950),[21] reference and predication have lain at the heart of Strawson's work. How exactly do we discursively refer to particulars? How can your talk of *this* thing or *that* thing allow your companions to focus on the very same thing that holds your attention? Strawson's thought, summarily put, is that reference is always reference *within a spatio-temporal framework* and always made in relation to perceivable spatio-temporal items, material objects or persons. Examining our linguistic practices he claims, first and foremost, that we identifyingly refer to objects by picking them out when they are sensibly present. We do this through *demonstrative* expressions; *this* pen, *that* dog, the piece of cream *there* on the tip of your nose. Reference succeeds when the audience picks out the same item in their own field of experience; they focus on a particular *thing* that is being indicated by the speaker.

We do, of course, refer in other ways too. Sometimes the item is not ready to hand – the forgotten book lies untended by your bedside, maybe the colour red floats in the Platonic realm of forms. We refer to these things by *describing* them. The novel insight that Strawson offers in *Individuals* is that referring to an object by this means ultimately relies on positioning the item in relation to objects in one's own environment. It can do so more or less directly; we may talk of the owner of this pen, for instance, or the dessert whence the speck of cream came. More obliquely, one's reference to Aristotle, say, relies on thinking of the philosopher as being alive at a certain point in history, in a certain place, which stands in a determinate relation to the time and place you yourself currently occupy. The background condition for reference is the supposition that one exists within a spatio-temporal framework among a variety of material objects.

Whether or not you agree with Strawson's analysis, his account of reference allows us to get a sense of the structure of his method. He looks to our ways of talking and finds therein certain items achieving a particular prominence. We cannot refer, he says, without presupposing material objects inhabiting a spatio-temporal framework. The success of reference depends on demonstrative expressions that pick out these entities. We might refer to qualities – the colour red – but to do so we must first pick out fire-engines and steaks and roses and so forth. Arguing along these lines he claims that publicly observable material objects are *referentially basic*. They are thus grammatically important in a way that other entities (like qualities) are not. Having stated this, he effects the descriptivist shift: if material objects are necessarily privileged in our linguistic procedures this is a consequence of our experiential scheme. We experience reality in a certain way. And the reason *why* we structure the world thus – thinking first of material objects and then, for example, of their properties – is because (he says) that reflects the way the world really is.

This is an impressionistic sketch. Still, it presents some of the characteristic features of descriptive metaphysics – among them, a focus on *language*, a focus on ontological *order* and a focus on *systematicity*. These are traits that Wiggins inherits, along with their defects.

### (i) Language

First of all, Strawson's method is a form of *linguistic analysis*. His immediate intellectual forebears are Russell and Moore and those of the 'ordinary language' school – but this technique for examining conceptual structures is, of course, not original to them. In using grammatical distinctions as a guide to conceptual ones, Strawson positions himself in a philosophical tradition that reaches back to, and beyond, Aristotle, who arrived at his categories by distinguishing the different questions that may be asked of a subject. Wiggins positions himself similarly, specifically in relation to the method of the *Categories*.

> If somebody claims of something named or unnamed that it moves, or runs or is white, he is liable to be asked the question by which Aristotle sought to define the category of substance: *what is it* that moves (or runs or is white)?[22]

Thus runs the first sentence of the first chapter of *Sameness and Substance Renewed* – and the refrain recurs throughout Wiggins's texts. His sortal theory of identity and individuation is motivated by the way we talk about *sameness*, and his claims about personhood emerge, in large part, from our *use* of the term 'person' (the target of the critique in Chapter 5, below). Like Strawson, like the ordinary language philosophers, like Aristotle, Wiggins aims to unearth the conceptual foundations that support our thoughts by looking to language. This strategy, as we will see, is not above reproach.

### (ii) Ontological order

We also find, in Strawson's discussion, a particular focus on *order*. He sees material objects, inhabitants of a spatio-temporal world, to be referentially 'basic'. When conversation turns to more diaphanous subjects – to colours, or numbers or the taste of butter – our reference to these things, Strawson says, *presupposes* reference to material objects. The everyday physical items which we can touch and track are grammatically more *fundamental* than the properties we see to inhere in them. Moreover, he takes this feature of our language to stand as a structural feature of our thought – and thereby, a structural feature of reality.

That reality has a structure might initially seem a bewildering idea. Is it not simply there? Are the items contained within it not all equally real? There are some philosophers who think that they are (these views will be discussed below), but there are reasons for thinking otherwise. Consider, for example, the existence of this book. Whatever you make think of its contents, few would want to deny that it *exists*. Nor would we want to deny the existence of cars or people, or the state of China, or depression, or the number '2'. But do all of these things exist in the same way? An electronic version of this text, and printed matter of the same, might seem to exist in different ways (one will burn easily, the other less so). It would be dreadfully confusing for mathematicians if you denied the existence of the number '2' – but asked to point to it, to hold it up to count it in an inventory of worldly objects, we will have problems. This thought, that there are different ways of being, is not accepted by metaphysicians *tout court*, but it certainly seems to characterize Wiggins's work. It is a central concern in the chapters that follow.

In Strawson, as in Aristotle and Kant, we find an interest in the architecture of reality. Seeing some entities to be linguistically basic, and thus conceptually basic, Strawson holds that they are somehow *metaphysically* basic. To resort to a more ancient idiom, some items are ontologically *prior*, and some, *posterior*. On this interpretation, descriptive metaphysics is concerned with clarifying the relationship between the items we pick out – their dependencies and interdependencies.[23] Susan Haack, in 'Descriptive and Revisionary Metaphysics', provides a helpfully concise statement of Strawson's position here:

> Ontological priority is defined in terms of our capacity to pick out and talk about things; [he] suggests that identifiability of at least some individuals of a given kind is necessary for the inclusion of that kind in our ontology, and

thus connects identifiability with ontological commitment, and identifiability-dependence with priority of ontological commitment.[24]

Does Wiggins (like Strawson) think that some entities *depend* on others? Are there metaphysically fundamental things that undergird the less resolute beings in the world? The answer – to be fleshed out in Chapters 3 and 6 – is yes ... and no. Wiggins recognizes this Strawsonian relation of 'ontological dependence', but he simultaneously countenances *causal* dependence. They might flow against each other, but Wiggins takes them both to characterize determinate features of reality.

### (iii)  Systematicity

Think back to Strawson's discussion of reference. He is not interested in a single isolated concept but in the whole conceptual framework. He is not concerned with reference alone, but with how this procedure figures in the larger picture of thought and language. According to the descriptivist, metaphysics cannot be conducted with a microscope; one must have a sense of the broader structure in order to understand the links between the constituent parts. In holding this, and advancing what he calls 'connective analysis',[25] Strawson positions himself in direct opposition to the 'piecemeal' attitude symptomatic of the high Analytic tradition,[26] articulated by Russell in *Mysticism and Logic*:

> A scientific philosophy such as I wish to recommend will be piecemeal and tentative like other sciences.
> To build up systems of the world, like Heine's German professors who knit together fragments of life and made an intelligible system out of them, is not, I believe, any more feasible than the discovery of the philosopher's stone. What is feasible is the understanding of general forms, and the division of traditional problems into a number of separate and less baffling questions. 'Divide and conquer' is the maxim of success here as elsewhere.[27]

This 'scientific' – not to say military – method, of separating out philosophical issues into 'manageable chunks' remains, as Glock notes, a prominent trend in Anglophone philosophy.[28] It is this approach that Strawson, in *Analysis and Metaphysics*, explicitly opposes.[29] There, he argues against what he sees to be the 'reductive or atomistic model' that underlies much of the practice of Analytic philosophy.[30] He opposes the dismantling form of analysis which strives towards 'a clear grasp of complex meanings by reducing them, without remainder', to isolated elements of meaning.[31] Strawson writes:

> Let us abandon the notion of perfect [atomistic] simplicity in concepts; let us abandon the notion that analysis must always be in the direction of

greater simplicity. Let us imagine, instead, the model of an elaborate network, a system of connected items, concepts, such that the function of each item, each concept, could, from the philosophical point of view, be properly understood only by grasping its connections with others.[32]

Despite the moratorium declared by Ryle, Strawson is a system-builder. So much should be expected from a philosopher who takes as his starting point the bountiful and endlessly expanding verdancy of natural language. And, significantly, the same is true for David Wiggins, who in his preamble to *Sameness and Substance Renewed* emphasizes how closely intertwined the notions of *identity*, *individuation* and *substance*, actually are, and how in examining one, it is necessary to examine them all and more.[33]

> What we have here to confront is a whole skein of connected practices. These practices are intertwined with one another.[34]

The importance of this in understanding Wiggins's sortal theory of identity and individuation is emphasized in Chapter 2. For now it will suffice to note that, like Strawson, he finds worth in trying to understand the shape of a network and not simply the points that constitute it.

*   *   *

This, then, is the Strawsonian frame on which Wiggins's philosophy hangs. His focus is the metaphysical relation of *identity*; his aim is to understand what makes an early object, a *substance*, the *same* as some later object. How, for example, are we to grasp the relation between the baby that becomes the woman, that becomes the mother, that becomes the patient in a vegetative state? Wiggins's response is at first blush simple and persuasive. We are, in our everyday lives, constantly confronted with identity questions and in practice, he says, we find it straightforward to answer them.[35] We have no trouble picking objects out – children, say – and observing their various exploits and adventures. There are everyday procedures we all engage in when we single out dogs or tea-cups, and which a child engages in when, for example, she learns that this cat is *her* cat, that it is the same cat as the kitten she was given and so on. We already have at our disposal a way of ruling on identity; so he suggests that in order to answer the more puzzling questions – those offered up by the personal identity debate – we can turn to this pre-theoretical method of navigating the world, and ground our answers upon that.[36]

This, then, is the methodological programme that underpins the work in *Sameness and Substance Renewed*:

> Let the philosopher elucidate *same, identical, substance, change, persist*, etc., directly and from within the same practices as those than an ordinary untheoretical human being is initiated into … let him shadow the practical

commerce between things singled out and thinkers who find their way around the world by singling out places and objects – singling out one another.[37]

## 3  Heritable defects

Not all inherited traits are useful. Some, indeed, are actively harmful. This section discusses some of the weaknesses found in the descriptivist account: its reliance on *conceptual invariance*, the apparent *failure to connect to reality* and its *rivalry with revisionary/scientific models*. Wiggins has the means to adapt to these shortcomings and the sub-sections below present various responses to them. The hope is not to construct a watertight defence against all related concerns; it is simply to identify certain theoretical pressures and to show that Wiggins's replies are consistent and part of a subtle and sophisticated response to such worries.

### (i)  Conceptual invariance, and a shift from linguistic analysis

Consider, first, the descriptivist's style of linguistic analysis and her reliance on the thought that humans, across all cultures, share a 'conceptual scheme' (a universal 'filing system', as we had it above). For Strawson, a metaphysician can articulate her metaphysical system by looking to the structure of experience; but, at base, are our experiences all so similarly structured? And if there is more than one scheme, which one is 'ours'? Which one should be our guide? Strawson himself notes that most concepts are culture-bound and temporary. These, surely, cannot be the basis for a metaphysical inquiry which aims to describe reality in *non-local* terms. Strawson and those who follow him must hold that 'there is a ... central core of human thinking which has no history ... there are categories and concepts which, in their most fundamental character, change not at all.'[38] Unfortunately, as Haack points out, Strawson does not actually tell us *which* concepts have 'no history'.

The massive grammatical variation we find in natural languages suggests that different linguistic communities may possess *different* conceptual schemes. Strawson's *Individuals* focuses on our modes of reference and predication as indications of the ontological priority of material objects – and this is a move that E.A. Burtt[39] and Tsu-Lin Mei[40] (among others[41]) have shown to be problematic. By analysing Chinese and Native American traditions, Burtt and Mei argue independently that the human 'conceptual scheme' which Strawson seeks to reify is local and temporary; the grammar of subject and predicate that pervades Latinate languages and Greek, and which Strawson takes to represent a central element of the human conceptual framework, is found to be absent in certain non-Latinate languages (notably some Sino-Tibetan languages and Quechua). Thus Burtt and Mei each claim that these analyses cannot provide a suitable foundation for metaphysical inquiry. (Even though it lies outside the official ambit of this book, it is important to note that these critiques emphasize how

Analytic metaphysics can function as a site of – conscious or unconscious – political activity.[42] If conceptual invariance fails to hold, the descriptivist is guilty of a form of metaphysical imperialism. Strawson's project reifies specifically Western forms of thought. Can it be that reality itself has the same structure as an Englishman's mind?[43] Surely not.)

In Chapter 5 it is shown how Wiggins is vulnerable to this line of argument. His Human Being Theory is grounded, in part, in the particular interests of the English language and the usage of the term 'person'. In doing so it ekes a metaphysical conclusion out of a culturally-specific linguistic artefact. While the conceptual analysis practiced by philosophers like Ryle and Ayer is focused on ironing out inconsistencies in philosophical discourse, Wiggins (following Strawson) aspires to make claims about reality. In doing so he asks of linguistic analysis a stability and neutrality it might find hard to maintain.

Nevertheless, Burtt and Mei's objection is not an insurmountable one for Strawson. His project does not stand or fall with the existence of a universal grammar, nor is the fact that different languages posit different entities by itself enough to undermine the conceptual invariance thesis. Rather, this fact causes us to question the centrality of linguistic/grammatical analysis to metaphysical inquiry. The descriptivist may yet claim that there are fundamental categories and concepts without looking to the nuances of our languages to determine them.

But it seems dangerously optimistic to think that, despite the huge variation in experience and the plurality of phenomenologies grounded in wildly different material circumstances, humans all conceive of the world in the same way. Turning to the rich work done on the phenomenology of pregnancy,[44] one gets a sense of the conceptual distances that lie between us. This is before we start thinking about cultural variance, through both space (between contemporary Americans, Japanese and South Africans), and time (between, say, twenty-first century Italians and pre-modern Etruscans). Is there not something overweening in the attempt to flatten these multifarious perspectives?

The strongest argument for conceptual invariance is not, I think, an argument but a hope. Though we may misunderstand one another, we are on the whole eminently capable of communication. If there is will on both sides, and enough time and resources, humans seem to be able to reach a common understanding. We may speak different languages, and enjoy different backgrounds, but we have a capacity for interpretation. Even in the direst circumstances there is always the possibility of accord. This seems to be a much more stable political resting place than faith in universal grammar. All people can, in theory, communicate with one another, and this is because they pick out the same items in the world. The capacity for interpretation suggests that there is a common framework in which we share (such thoughts are expanded in Chapter 4, below).

I say we share a common conceptual *framework* because the notion of a 'conceptual scheme', to which descriptive-style philosophers typically defer, is powerfully loaded. Even though it has its own connotations, I find 'framework' to be a more neutral term. The conceptual scheme is that which surfaces in our grammar. In contemporary debates about such things, different languages may be said to support different schemes.[45, 46] It may even be that users who *share* a language bring different schemes to bear. Suppose two people, for example, endorse opposing scientific models (Einsteinian versus Newtonian).[47] There are, as a result, an awful lot of schemes. A speaker of Cantonese may schematize the world differently from a French speaker.[48] An individual who believes the earth is flat may have a different scheme from one who believes it to be an oblate spheroid.[49] On this understanding, the standard view of schemes offers an embarrassment of riches.

To cool our blushes, let us talk then of 'frameworks'. A framework will sustain different schemes. Writing of the *pre-theoretical* framework, I have in mind the structure that allows humans to communicate (at least, in principle), makes disagreements between language users possible and allows them – one hopes – to overcome them. The *scientific* framework, on the other hand, is the description of reality that appears if we abide by the system of rules to which scientific inquiry makes itself answerable; its methodological programs, its commitment to consistency, explanation and predictability.[50] The scientific framework is the picture that scientists are working towards, in constantly revising their theories.[51]

These two frameworks – the *scientific* and the *pre-theoretical* – are at the fore of the discussion in this book, but there may be others besides. Implicit in the description of the pre-theoretical picture is the thought that it is *our* picture. It outlines the concepts *humans* must possess to be able to navigate the world thus and so. One need not think of extra-terrestrial life-forms to wonder if there are other such frameworks, possessed by beings of dramatically different physiologies. Fleas, perhaps, might not grasp concepts, but dolphins? Apes? These creatures seem very much to see the world in different ways from the way we do – ways one might well think realize different pre-theoretical frameworks. (Insofar as there is a problem about anthropocentricism, it will be discussed in due course.)

Wiggins may well be open to this terminological modification. It is not, I think, an accident that he tends not to talk of schemes. What I am proposing may help him respond to Burtt and Mei's concerns, but the focus on frameworks rather than schemes provokes its own difficulties. Part of the appeal of Strawson's approach and the analysis of grammar is that it presents an easily readable resource for understanding the interactions between concepts. Languages are public. 'They can be dissected and discussed in the open air' – so writes Michael P. Lynch, in contrasting the 'ordinary language' method to its predecessors. '[Languages] are therefore much more palatable to the scientific mind than murky Kantian "categories"'.[52] This was one of the initial attractions of Strawson's approach. It drew metaphysics out of the murk. How successful Wiggins is at deriving his philosophy from these seemingly deeper features, rather than from through the direct analysis of language, remains to be seen.

### (ii)  Conceptualism, and a failure to connect

The descriptivist philosopher must defend conceptual invariance and find a reliable method for accessing the pre-theoretical framework (where linguistic analysis is found to fall short). Questions still remain. The sceptic will ask by what right one can simply assume that our pre-theoretical framework connects with reality in the requisite way. Why exactly do Strawson and his intellectual descendants think that the structure of experience can correspond to the structure of reality? That we *feel* that certain features of the world are non-contingent – the primacy of everyday material objects, for instance – does not necessarily mean that they really are. Descriptive metaphysics may yield insight, not into reality, but into our forms of description of reality. It is a point put clearly and poetically by P.M.S. Hacker:

> Strawson ... saved the letter of traditional metaphysics, but abandoned its spirit. Descriptive metaphysics is distinctive, and unlike any other philosophical endeavour, in so far as it strives to disclose the most general forms of connectedness that permeate our conceptual scheme and to reveal the conceptual involvements of the most general kinds of speech functions that characterize our use of language – indeed, not only of *our* language and *our* conceptual scheme but of any language and any conceptual scheme in which certain kinds of distinction are drawn and certain kinds of speech acts are performed. But it is also like other philosophical endeavours within the field of connective analysis and unlike the aspirations of traditional metaphysics. It yields no knowledge of reality, let alone insight into the necessary structure of reality – but only insight into the forms and structures of our thought about reality. It might indeed be said to be the legitimate heir to what used to be conceived of as metaphysics, but a dethroned heir, deprived of the ancestral crown and orb. It is metaphysics without its nimbus.[53]

Has Wiggins made efforts to restore this nimbus – or is his project simply an attempt to describe the limits of our experiential structures? It must be said that, if he aims to stretch beyond conceptual analysis, his arguments for how he does so are somewhat obscure. His primary concern, here, is not that descriptive accounts may fall shy of reality but that they might be open to complaints of 'conceptualism'. So let us start with these complaints and see where they take us.

Wiggins is not immediately concerned with whether or not his metaphysics stands on a par with the projects of 'traditional' metaphysics; rather, his worries revolve around accusations of *idealism*.[54] If metaphysical inquiry is grounded in conceptual analysis is this because human categories somehow *shape* reality? This is the 'conceptualist' picture he describes wherein the echoes of Kantian idealism are found.[55] This is this echo that Wiggins seeks to silence in developing his so-called 'conceptualist-realism'.

In caricature at least, the conceptualist who Wiggins opposes holds that the macroscopic order we find in the world is a product of human cognition. We look around and see all manner of stuff: lumps of metal and wood, and flesh and bone. We take this stuff to constitute mountains and chairs, and dogs and cats, and so on and on. For the conceptualist, there are such things *because* we conceptualize them as units;[56] i.e. the order we find in the world is *produced*, in some way, by our conceptual framework. As Wiggins notes, the idealist extremes of the position are illustrated in Leszek Kolakowski's *Towards a Marxist Humanism*:

> The picture of reality sketched by everyday perception and by scientific thinking is a kind of human creation (not imitation) since both the linguistic and the scientific divisions of the world into particular objects arise from man's practical needs. In this sense the world's products must be considered artificial. In this world the sun and stars exist because man is able to make them *his* objects, differentiated in material and conceived as 'corporeal individuals'. In abstract, nothing prevents us from dissecting surrounding material into fragments constructed in a manner completely different from what we are used to. Thus, speaking more simply we could build a world where there ... might be, for example, such objects as 'half a horse and a piece of river', 'my ear and the moon', and other similar products of a surrealist imagination.[57]

Certainly, if one is a conceptualist, one is likely to be a descriptivist in method. How can the conceptualist answer questions of continuant identity otherwise than by turning to the everyday practices and pre-theoretical schema of which she takes those continuants to be products? The question is whether the converse holds. Is the descriptive style of philosophy necessarily committed to this sort of conceptualism? Briefly put, Wiggins's answer is *no*. One of the lasting innovations of his project is to detach the descriptive method from all Kolakowskian forms of conceptualism. The key is his so-called 'conceptualist-realism', a form of realism that states that our conceptual framework maps reality, not because it constructs it, but because it develops in reciprocity with it.

All sorts of objections can be raised against the kind of conceptualism we find in Kolakowski's work. Many are articulated by Wiggins himself.[58] Chief among them is the thought that conceptualism is committed to *bare substrata*[59] – entities, which may be conceptualized in radically different ways, by different beings, and assigned radically different principles of existence and persistence. How might we realistically conceive of a quality-less substrate that supports inconsistent properties? Wiggins's answer is simple: we cannot. To do so would be to ignore the exigencies of logical identity. Moreover, conceptualism is associated with the unlovely (and painfully abstruse) notion of *haecceity*.[60] Wiggins finds it inimical to any worthwhile essentialist programme.[61] More fundamentally, in its idealist leanings, Kolakowski's conceptualism is profoundly at odds with the powerful and persuasive *realist* thought that the objects we find around us – people, trees, lakes and so on – exist quite independently of how human beings conceive of them.[62] The macro-order we divine around us exists in its own right.[63] It is this thought that animates metaphysical position we find in *Sameness and Substance*.[64]

Despite recoiling from it, Wiggins does not reject everything he discovers in Kolakowski's work. For while he sees dangers in a view that veers so close to idealism, he is impressed by the thought that reality does not *separate itself out for us*. We bring structure to the table, so to speak, the table does not simply bring structure to itself. Here he positions himself in opposition to what he calls the 'myth of the *self-differentiating object*'.[65] The world, he says, does not offer itself up to us fully partitioned, clearly divided. It does not frame itself – we frame it – but its boundaries are no less real because of this. This 'conceptualist-realism'[66] is caught, neatly, when he recasts Arthur Eddington's analogy of the net:[67]

> Our claim was only that what sortal concepts we bring to bear upon experience determines what we can find there – just as the size and mesh of a net determine, not what fish are in the sea, but which ones we shall catch.[68]

This Strawsonian approach[69] allows Wiggins to hold both a form of conceptualism *and* a form of realism. The two are not incompatible. Our conceptual framework is important, but not because it constructs the world. It is important because it allows us to pick on particular features of it. Other features may be picked out in other frameworks, such as the scientific one (I return to this later in relation to his broad-minded realist picture). In *Sameness and Substance Renewed*, Wiggins writes:

> For someone to single out a leaf or a horse or a sun or a star, or whatever it is, that which he singles out must have the right principle of individuation for a leaf or a horse or a sun or a star.... For to single out one of these things he must single *it* out. Such truisms would scarcely be worth writing down if philosophy were not driven from side to side here of the almost unnegotiable strait that divides the realist myth of the *self-differentiating object* (the object which announce itself as the very object it is to any mind, however passive or of whatever orientation) from the *substratum* myth that is the recurrent temptation of bad conceptualism. It is easy to scoff at *substratum*. It is less easy to escape the insidious idea that there can be the singling out in a place of a merely determinable space-occupier awaiting incongruent or discordant substantial determinations (individuatively inconsistent answers to the question of *what it is*). But no substance has been singled out at all until something makes it determinate *which* entity has been singled out; and for this to be determinate, there must be something in the singling out that makes it determinate which principle is the principle of individuation for the entity and under what family of individuatively concordant sortal concepts it is to be subsumed.[70]

There are a number of technical terms here that will be introduced in due course. At this point, my citation is simply intended to give an indication of Wiggins's conceptualist sympathies and to show how, *contra* what he sees to be the prevailing orthodoxy,[71] he takes conceptualism and realism to be, to some extent, compatible. Though he denies that the mind *constructs* reality – in the 'bad' conceptualist sense – he argues that it does still *construe* it in a certain way. This, then, is how he responds to the accusation of conceptualism. He accepts it. But construal is not construction. His sober conceptualism is, he says, fully consonant with realism; they are not exclusive.

Sceptics, however, are nothing if not persistent. They will point to the complaint with which we started this section. We were not concerned – they will say – with *how* Wiggins is conducting metaphysical inquiry; we are concerned that he is doing metaphysics at all. Even if conceptualism and realism are, in principle, compatible, what reason does he give to suppose that our concepts do in fact attach to structural features of reality? It is strange, given the apparent trenchancy of this objection, that Wiggins avoids meeting it head on. On one level, his response seems to be deflationary. He treats the connection between thought and reality as a given (or at least a connection to be defended elsewhere). Thus he writes in the introduction to his latest collection:

> The general background ... is a simple philosophical realism: the claim (to be defended somewhere else in philosophy) that what we confront and try to describe – and that whose workings we seek earnestly to understand – is a mind-independent reality. We are not cut off from all success in that endeavour. This is to say that, even where the observer contributes a particular perspective or slant upon that reality, the reality that is revealed is not simply by that token a construct or creation of the knowing mind or the effort of cognition.[72]

Like the sceptic, we might find this more than a little unsatisfying. Why is this the general background – and how can he assume it? Fortunately, Wiggins seems to have another, more critically engaged, response. This issues from his discussion of conceptualist-realism. Therein, as mentioned, he draws attention to our *reciprocal* involvement with the world. We are engaged in a constant give-and-take with our environment. Wiggins's method proceeds by tracking our 'practical commerce',[73] and our 'untheoretical business',[74] with the world. Sometimes it meets our expectations and sometimes it doesn't. We adapt and refine our actions accordingly. That is to say, the fisherman tightens and loosens her net depending on her *successes*.

Wiggins claims that our conceptual framework is 'open to the world'.[75] To think that our pre-theoretical world-view is a 'human creation (not imitation)' ignores, he says, how it has developed through *a reciprocal engagement* with reality.[76] Certainly, when we pick out items in our environment, we conceive of them in specific ways, bound as we are by our human nature with our particular perceptual abilities. But we also *interact* with reality and, in interacting with it, we can discover more about it. We can modulate our conception of it to match more closely with how it must *actually be*.[77] This is a process we are constantly engaged in.[78]

> [E]ven though horses, leaves, sun and stars are not inventions or artefacts, still, if such things as horses, leaves, sun and stars were to be singled out in experience at all so as to become the objects of thought, then some scheme

had to be fashioned or formed, in the back and forth process between recurrent traits in nature and would-be cognitive conceptions of these traits, that made it possible for them to be picked out.[79]

The recognition of this reciprocity, between what Quine calls the *ontological* and the *ideological*,[80] is central to Wiggins's methodological programme.[81] Crudely put, he thinks that once we attend to the conceptual framework that underlies our everyday practices, we can see that it *can* give us a guide to how reality is structured. It was brokered in *response* to reality, not enforced upon it nor irrelevant to it.[82] The efficacy of this reciprocal engagement is central to our lives. The basic notions of identity and persistence sustain our survival ('persistence through change is not make-believe. No sensible inquiry could abandon a datum so fundamental or so deeply entrenched.'[83]).

In line with the comments in the section above, Wiggins can say that his focus is not upon surface-level grammar, about which there may be disagreement, but rather the framework that allows for that disagreement. This is the framework we cannot help but work within; it is essential to our navigation of the world. Think, he will say, of the dislocation that would result if we were to try to treat the world, not as containing material objects persisting through time, but as composed of four-dimensional entities[84] or processual beings. How much more difficult it would be to talk to, or smile at, or love someone if you no longer saw a single, unified being, but a constantly fluctuating miasma of mereological simples? How would one survive – find food and consume it, find love and enjoy it – if one could not pick out and trace determinate objects through space and time? Human life rests upon our ability to read reality, to conceptualize it in a way that tracks it. If the net did not work, if it failed to catch fish, the fisherman would go hungry. If our conceptual framework did not correspond to certain structural features of reality, then – the thought goes – we would find ourselves adrift, unable to cope with the experiential stream.

\* \* \*

We see in Wiggins's work a sort of bridge to span the gap between our thoughts and the world to which they are directed. On the interpretation offered above, he is not concerned with *local* conceptual schemes, as revealed by grammatical architectures. Rather, he takes as his subject the system of thought that sustains interpretation, which allows us to move through the world in the characteristic way that we do. We get about, we communicate. So we must, in the end, be getting something right. Nor need this be because we are moulding the world to our will or forcing a structure upon it. It is not necessary to be a conceptualist or an idealist to hold Wiggins's position. One need only think, as he does, that we are reciprocally engaged with our environment. Sometimes our expectations are frustrated, sometimes they are met. It is not a case of simply wishing the world to be a certain way and finding it so – we work hard to understand it and are rewarded by being able to predict it, and to live within it.

### (iii)  Scientific rivalry and Wiggins's verdant realism

> Yes; there are duplicates of every object about me – two tables, two chairs, two pens.... One of them has been familiar to me from earliest years. It is a commonplace object of that environment which I call the world.... It has extension; it is comparatively permanent; it is coloured; above all it is substantial.... Table No. 2 is my scientific table.... My scientific table is mostly emptiness. Sparsely scattered in that emptiness are numerous electric charges rushing about with great speed; but their combined bulk amounts to less than a billionth of the bulk of the table itself.[85]

Let us assume the viability of conceptualist-realism, and forge onwards. Who is then to say that the descriptive approach is the most *accurate* method for mapping metaphysical structure? Introspection may offer a *sense* of the shape of the world, but it is a dark glass and perhaps there are better ways. The natural sciences are often taken to be a more accurate guide to the nature of nature. Maybe, by turning to physics, we can achieve a critically purified picture of reality, which escapes the descriptivist's vulgar, anthropocentric world-view.

The quotation above, from Arthur Eddington's famous colloquy on the two coincident tables, illustrates the contrast between the 'purified' and the 'vulgar' conceptions of the world. No one, of course, would want to deny the existence of Table No. 1 but Table No. 2 – the scientific table – appears to more accurately reflect the make-up of reality. While the common-sense table is apprehensible to humans it might, one may think, escape detection by life-forms with different perceptual abilities. The scientific table, however, is described from an 'objective' standpoint, a 'view from nowhere',[86] it represents an 'absolute conception'[87] of the world and will not brook these kinds of disagreement. So the thought goes. Scientific inquiry offers clearer, more precise descriptions and thus stands as a better starting point for metaphysical exploration. This is the tack that revisionists, in Strawson's sense, will want to take.

Moreover, they may say, science stands as a rigorous system of knowledge, produced according to strict rules and principles. It presents us with a vast repository of information about the world, the result of extensive and sophisticated testing. Scientists must abide by certain modes of practice; every assumption, every variable must be quantified and analysed. The same cannot be said of 'common-sense' or whatever measure the descriptivist uses to map out our pre-theoretical framework. There are no rules in place to refine pre-theory.

Let us take this last concern first. Our lives are undoubtedly improved by the results of scientific endeavours. Nevertheless, small successes are achieved every day which are better attributed to our pre-theoretical framework. This morning I found my tooth brush; I managed to walk through the front door of my house; I have found my lunch and have eaten it. None of this was accidental. The pre-theoretical framework, which allows me to pick out objects and to trace them through time, underpins all of these activities. Nor should we be surprised by these successes. The science-led metaphysicians will hopefully be impressed by the fact that the method by which I pick my way through the world is the outcome of millennia of human evolution and the rigorous requirements of natural selection, which have refined my modes of individuation. Our survival, as individuals and as a species, stands as lasting proof of the efficacy of our pre-theoretical framework.

Another point the descriptivist can make is that scientific inquiry is *grounded* in pre-theory. The questions that the scientist asks – about tables, and all those other humdrum things – are necessarily positioned within our everyday conceptual framework. Before scientists from alien cultures can start agreeing about the nature of reality they must first see each other as beings who can engage in joint projects. These are the background conditions for the scientific enterprise. That is, in responding to the worries above, Wiggins and his supporters can raise similar concerns about the revisionist's own methodological program. The revisionary metaphysician privileges science over everyday experience because it can – supposedly – reach beyond our limited anthropocentric concerns. But Wiggins may want to reply that since science is a *human* practice it must be grounded in our pre-theoretical framework, it cannot displace it.

Susan Haack finds this 'ambitious descriptivism' in Strawson,[88] and a version of it is present also in the quotation from Kolakowski (given above). Kolakowski writes that 'scientific thinking' is a human creation; he holds that the linguistic *and* the scientific divisions of the world 'arise from man's practical need'.[89] Of course, Wiggins's modest conceptualism is different from Kolakowski's, but there are indications in his texts of similar thoughts. They emerge, for instance, in the discussion of the 'alien' language of 'four dimensionalism',[90] where Wiggins writes:

> At one and the same time, how can we deny ordinary substances their status as proper continuants, insist that ordinary substances are really *constructs*, yet lean shamelessly upon our ordinary understanding of substances when we come to specify that from which these constructs are to be seen as constructed or assembled?[91]

Wiggins does not deny that science posits entities that fail to register in our everyday experience – microphysical particles, say, or four-dimensional beings – but remarks that the practices that issue in their discovery depend themselves on our pre-theoretical framework. To call those scientists who seek to do without the pre-theoretical 'shameless' is more than a moral censure. They are not only forsaking the ordinary world, which gives rise to the ordinary concerns that motivate scientific explanation; they do so on pain of inconsistency. Our idea of a substance, Wiggins says, of a material continuant that undergoes changes and weathers the world, sustains the less familiar notions of 'time-slices' and 'mereological simples', and to disavow it is to saw the branch from underneath one's feet. On these grounds, therefore, one may question the revisionist's impulse to privilege the entities of the 'older' ontology with those found in the 'newer' one.[92]

Unlike Strawson, Wiggins does not take the descriptivist's articulation of reality or the entities that it describes to be in any way 'prior' to those posited by science.[93] Wiggins is not claiming that everyday objects are *more real* than the stuff from which they are made. The suggestion contained in the quotations above is simply that the converse claim should be resisted. Physics describes new entities – atoms, quarks, etc. – but they should not be taken to *downgrade* or *discredit* the entities of the older, pre-theoretical ontology. Even when the older entities are discarded for certain explanatory purposes, it need not be the case that scientific revisions should legislate metaphysical ones. The general point here is that Wiggins has reasons to question the viability of the *revisionary* use of science, with respect to our pre-theoretical framework. And this encourages the thought that the items of everyday experience are at least as well grounded as those of the scientific scheme. Eddington's tables – No. 1 and No. 2 – are *equally real*.

Here, the science-minded philosopher will persist. The two pictures are not only different, she will say, they are in *conflict*. The common-sense table is 'substantial' and solid; the scientific table is nearly all empty space, neither substantial nor solid. The scientific and pre-theoretical frameworks are describing one item in radically different ways. Is the table solid or not? It cannot, surely, be both. How can Wiggins respond to this? Fortunately, the conflict disappears if one attends more closely to the response given above. Science picks out some entities, pre-theory picks out others. The conflict rests on the assumption that the two frameworks are talking about *the same thing*, in incompatible ways. Wiggins, like Susan Stebbing[94] and Amie Thomasson,[95] can deny that Table No. 1 and Table No. 2 are the same thing, spoken about in incompatible ways. Science picks out a mass of atoms, *we* – you and I and other ordinary human beings – pick out a table.[96] We cannot pick out the atoms (we cannot even see them). In a passage which anticipates the picture to follow, Wiggins writes:

> The older ontology may yet be *cotenable* with the more theoretical conception. Contrasting the actual discrediting of entities of some kind, palpable or impalpable, with the discovering of new entities at the atomic or subatomic level, let us not conceive the latter as determining the level to which everything else must be reduced (in the serious sense of 'reduce'), even if this is the level at which macroscopic events are promised certain sorts of explanation.[97]

\* \* \*

We have wandered quite far from the main paths; the flora is unfamiliar now, the ground untrodden. This is meta-metaphysics, where one asks not just questions about existence, identity and substance, but questions about those questions. This landscape, though visible in Wiggins's work, is some distance from his main concerns. But it is the scenery against which everything else unfolds.

On the reading offered above, Wiggins thinks there is a mind-independent world, which we strive to understand. He thinks too that reality contains *various types of entities*. There are the entities captured by our pre-theoretical framework, persisting material beings – or *substances*, as he says – and there are entities found in the scientific framework, quarks and fields and such. One gets a sense, from the quotations, that he thinks other entities exist besides – four-dimensional things, for example, that have spatial *and* temporal parts (and there are others as well, like the 'concrete universals', which will be discussed in Chapters 2 and 9, and the 'processes' described in Chapter 7).

Moreover, from what has been said, and from what is to follow, Wiggins can be interpreted as holding that, in picking out these different entities, these frameworks – the scientific, the pre-theoretical – make no claims to be *complete* and *exhaustive*. They are different nets, so to speak, which catch different fish. You may sweep with one and still find more creatures in the sea. In claiming this, Wiggins stands in opposition to what we may term 'hardcore realism'[98] or 'metaphysical realism'. Wiggins makes this explicit in the introduction to his most recent collection, writing:

> Philosophical realism needs to be distinguished from another kind of realism which Hilary Putnam calls *metaphysical* realism. This is the doubtful contention that any 'given … system of things can be described in exactly one way, *if* the description is complete and correct.… The [correct] way … fix[es] exactly one ontology and one ideology (in Quine's sense of those words) … [it fixes] exactly one domain of individuals and one domain of [properties] of those individuals'. I concur with Putnam's rejection of these claims and note that the "conceptual realism" proposed in Chapter Five of *S&S* and *S&SR* was all of a piece with a similar rejection.[99]

There is no single, exact ontology, described from a single, exact perspective. There are different angles on reality and each throws light on different aspects. It might then seem as if Wiggins is endorsing a form of metaphysical *pluralism*. The affiliation with Putnam – whose 'conceptual pluralism' is a powerful exemplar of that position – could support this reading.[100] Yet, as elsewhere, it is important to be clear about the meaning of the terms before attributing them. On the one hand, pluralism might be taken simply to be the endorsement of a plurality of metaphysical perspectives. On the other, it might be the endorsement of a plurality of potentially *incompatible* metaphysical perspectives. This latter reading is found in Michael P. Lynch's *Truth in Context*, which also adverts to one of the most common criticisms of metaphysical pluralism: that it stands as an affront to the *objectivity of truth*.[101] The different frameworks must describe reality in *conflicting* ways (as Stebbing, Thomasson and Wiggins all note). (One framework might, for instance, deny the existence of God, while the other might rely upon it. One might pick out chairs and tables while another might say there exists nothing more than mereological simples.)

Wiggins is not a member of the Lynch mob. Nowhere does he argue for the incompatibility of metaphysical pictures – indeed, to do so would be an offence to his sense of logical hygiene. He is a *realist* – but a broad-minded one.[102] Like Thomassson and Stebbing he denies that the scientific framework and the pre-theoretical framework can claim to provide a *true* and *complete* description of reality. His picture is more than a little similar to the 'promiscuous realism' of John Dupré (discussed in Chapters 6 and 7),[103] except that Wiggins is interested in both the reality of various entities *and* their metaphysical characters. There are many different things, and for Wiggins they exist in multifarious ways. His theory, then, is as he puts it, 'a theory embracing all sorts of other categories of being beside substance – categories of event, process, state, disposition, field, and heaven knows what else'.[104]

Things have become increasingly gestural, so let me at least gesture in a helpful direction. In reading Wiggins as an open-minded metaphysical realist, I am situating him in a tradition of thinkers, like Thomasson,[105] Kris McDaniel,[106] and Roman Ingarden,[107] who see metaphysical inquiry to be directed at a plurality of equally real, *orthogonal* systems. While Wiggins is interested in just one of these frameworks (our pre-theoretical one) he countenances the legitimacy of others. His focus is on the world relations picked out in our pre-theory but he recognizes that the scientific framework traces other world relations, which are no less real for being different – among them, *causal* links (discussed in Chapter 7).

\*    \*    \*

It is not altogether whimsical to say that a philosopher's temperament appears in the metaphors she uses. Metaphysicians talk of 'carving nature at the joints' or 'casting nets'. The former metaphor is noticeably loaded; it leads one to think that metaphysics aims towards a single, true description of reality.[108] The second – Eddington's – is not quite so bloodthirsty, nor so exclusive. Still, it carries strange connotations: that to study one's subject, it must first be caught (and killed?); that one may pursue one's ventures without any mind to the variety of life that broils beneath the surface.

Perhaps a different metaphor would encourage metaphysicians to see their project in different ways. Rather than butchers or hunters, maybe we can think of metaphysicians as inhabitants of a particularly bountiful ecological niche. The eco-system is, on the whole, well-balanced, and the different species of metaphysician draw various things from, and contribute other things to, their environment. Competition exists, yet the staggering abundance of reality nourishes metaphysicians of different species, without necessarily sustaining one over another.

## 4  Persisting problems

One can admire the exhaustive – if sometimes exhausting – rigour found in Analytic philosophy without wishing to emulate it. The sections above have presented a variety of problems for the descriptive metaphysician, along with a handful of responses. Some of these responses find fuller form in the chapters below – but in general, the hope here is not to run (or plod), through the pros and cons and counter-pros and counter-cons of the assorted permutations. That would require a much longer work than the present one, and one that would hardly be more interesting. The ambition here, rather, has been to map out the intellectual environment of which David Wiggins is a part and to identify the forces acting in this philosophical niche. In responding to the worries above, Wiggins positions himself as a broadminded metaphysical realist. Doing so he invites further criticisms – outlined now – some of which are to be expanded in due course, some to be skirted over.

First, one might worry that the verdancy of Wiggins's metaphysics threatens the principle of *ontological parsimony*. Where Quine had tastes for desert land-scapes,[109] Wiggins's picture is rich, fertile and full of life. And it would be troub-ling, certainly, if he admitted entities into his picture indiscriminately. But though he countenances multiple categories of being these categories are far from arbitrary. The mereological simples posited by science, for example, have a sound theoretical basis; quarks are the posits of sophisticated physical hypo-theses. Everyday creatures are similarly well-founded, by way of pre-theory; our mode of life depends on picking such things out, our continued survival stands as evidence for their existence. The appeal to ontological parsimony is, as Quine recognized, aesthetic in nature; Wiggins's tastes – in themselves, neither good nor bad – are for organic abundance.[110]

A second worry – about *dependence* – appears when we consider the relations between these different descriptions of reality. Sellars claimed that the manifest image and the scientific image purport to provide true and *complete* descriptions of reality and, doing so, stand in conflict. In contrast, Wiggins holds that the sci-entific and the pre-theoretical framework should make no claims to complete-ness. Moreover, he suggests that the two frameworks seem to *depend* on each other in certain ways. Yet if the two frameworks are joined by lines of depend-ence, might one be more *fundamental*? Might we *reduce* one to the other; might the entities of the manifest image supervene on those of the scientific image? This question is explored in greater depth in Chapters 7, 8 and 9 below. The pro-posal there is that while lines of dependence may exist, they exist in different forms (causal or ontological) flowing in different directions, and there is no obvious need to prioritise one over the other.

There is a third worry about *coincident entities*. Wiggins acknowledges the existence of multifarious beings – mereological simples, concrete universals, substances and more – and oftentimes these entities appear to materially coin-cide. How is this possible? How might one plausibly hold that, where you are currently sat, there is an organism, a biological artefact, a limb of a four-dimensional being, a concrete universal and an aggregate of mereological simples? All such things are made of matter; how can they be in exactly the same place at exactly the same time? Wiggins's 'is of constitution' indicates a sophisticated answer to this worry, and is fleshed out further in Chapter 2.

Through the various drafts of this book, the following, fourth question has risen to increasing prominence. It is a question about *correlation*. Wiggins sug-gests that we can sure our grasp on the persistence conditions of everyday organ-isms through biological inquiry. He thinks, that is, that we may talk about entities in one framework from the perspective of another. How though, can the biologist speak to the metaphysician about Aristotelian substances, for example? Is there a theoretical framework that matches pre-theory (if that is what the Aristotelian picture captures)? Is there a developed model of organisms, which maps exactly onto the kinds of creatures we talk to, and laugh with, and whose births we rejoice, and whose passing we mourn? There are, as will be discussed in Chapter 6, many theories about biological individuality – and none of them seem to correspond exactly to the everyday organism. A crucial thought is that different frameworks, fixing on different entities, might have limited explanatory reach. The biologist may be well-placed to discuss the individuals posited by biological theory, yet unable to offer insight into the persistence conditions of a human being.[111] The repercussions of this thought will be discussed, but rather than grapple with the bewildering complexities of orthogonal ontological systems I will offer a more modest response: irrespective of whether or not the human being, as a pre-theoretical entity, can be examined effectively through the scientific lens, we have other ways of investigating its persistence conditions. Moreover, we might find that the demand for these conditions is less pressing than originally supposed. These are the views essayed in the latter part of this book.

*  *  *

David Wiggins's work inhabits a very particular niche in Analytic metaphysics, and Analytic metaphysics itself is but a small region of Anglophone philosophy, and an even smaller one in the discipline more generally. His work does not, however, suffer for want of richness; the aim of this first chapter has been to show how it stands as a careful and detailed response to various environmental pressures, how it emerged in reaction to specific philosophical concerns and how it persists in a competitive intellectual microcosm. This is the setting – vital, busy and sometimes dangerous – for the discussions that follow.

## Notes

1 For a much less gestural account of the emergence of descriptive metaphysics, I recommend Peter Hacker's 'On Strawson's Rehabilitation of Metaphysics' (2003).
2 Ryle 1970: 10.
3 See, for example, Chalmers, Manley and Wasserman 2009.
4 See, for example, Hans-Johann Glock 'Strawson and Analytic Kantianism', in Glock 2003.
5 For a worthwhile introduction to, and discussion of, conceptual analysis, see Michael Beaney's 'Analysis' (2015).
6 Strawson was not, of course, the only figure to resist the anti-metaphysical attitude. Hacker cites, e.g. H.H. Price as another member of the resistance (in Hacker 2003).
7 Hacker 2003: 49.
8 Hacker 2003: 49.
9 Strawson 1959.
10 Strawson 1966.

11 Glock 2003: 15 (see also S. Körner 1966).
12 Glock 2003: 24.
13 Grundmann and Misselhorn 2003: 206.
14 Glock comments that:

> part I of *Individuals* elaborates a Kantian idea; namely, that our reference to objects depends on our capacity to identity and re-identify them, which in turn depends on the possibility of locating them within a single public and unified framework, the framework of the spatio-temporal world.
>
> (2003: 19)

15 Though I have been unable to secure a copy, I believe Kalyankumar Bagchi's *Descriptive Metaphysics and Phenomenology* (1980) is an extant case of a comparison. The fact that it is so difficult to find in itself speaks volumes (though the task of writing up said volumes must fall to someone other than me).
16 One author who seems to be encouraging a resurgence of interest in 'analytically minded' phenomenological discussion, is Amie Thomasson (see, for example, Thomasson 1999).
17 Strawson 1959: 9.
18 Strawson 1959: 9–11.
19 Mach 1905: 148.
20 Thomas Pradeu has pointed out to me that there is a way to *ally* science with descriptivist metaphysics. Consider, for example, neurologists, psychologists, and other such folk, who examine how we experience the world. I cannot dissent – but it is worth emphasizing that *these* scientific investigations will not be able to contradict our everyday thoughts about the world or offer any revision of them.
21 Strawson 1950.
22 Wiggins 2001: 21.
23 Hacker 2003: 49.
24 Haack 1978: 362.
25 Strawson 1992: 21.
26 Joll 2010: §iii.
27 Russell 1925: 109.
28 Glock 2003: 165.
29 Strawson 1992. For a clear articulation of Strawson's position here, see Byrne 2001.
30 Strawson 1992: 19, 21.
31 Strawson 1992: 17.
32 Strawson 1992: 19.
33 Wiggins 2001: 18.
34 Wiggins 2001: 19.
35 Wiggins 2001: 56.
36 Wiggins 2001: 2.
37 Wiggins 2001: 2.
38 Strawson 1959: 10.
39 Burtt 1953.
40 Mei 1961. See also Hacking 1968.
41 We find foreshadowing of this line of critique in Whitehead's *Concept of Nature* (Whitehead 1919, see Haack 1978 for commentary). (NB Whitehead's position – which appears *before* Strawson's – is notable too, in relation to the processual view he advocates; it will be discussed at the end of Chapter 7.).
42 Haslanger (2012) is an excellent guide to the hidden politics of Anglophone metaphysics.
43 Curiously, this brings to mind the Hebrewist attitude we find in seventeenth century metaphysics (and which some commentators find in Spinoza): the thought that Hebrew, as God's language, the language of the Chosen, is the linguistic framework

that can most accurately describe reality (a connection that Mogens Laerke has helped me make). The suppressed thought is that, in a manner of speaking, the English language (or at least, Latinate ones) provides the best means for discussing reality.

44  E.g. Louise Levesque-Lopman 1983, Francine Wynn 2002.

45  See, for example, Putnam, Thomasson and Button's discussions (in the works cited in the bibliography). Button, in *The Limits of Realism*, writes that 'Putnam has always maintained that the two languages can express different schemes even though they are intertranslatable' (2013: 212).

46  Lynch 1998: 35.

47  Button fleshes out this picture, and we find even more disagreements if we countenance the difference between 'optional' languages and 'natural' languages (as Jennifer Case does). See Button 2013: 213.

48  The representation of gender in linguistic structures may be different; the worries expressed by Luce Irigaray about the grammatical privileging of men over women (which we would surely want to deny reflect reality) might not appear in other languages.

49  Thomasson 2003: 143. As Thomasson sees it, our commonsensical conceptual scheme contains, for example, the thought that the earth is flat (as opposed to being an oblate spheroid). This may be a feature of a conceptual scheme but it is not, I think, a view we are obliged to hold in virtue of our everyday navigation of the world.

50  Lorraine Daston gives an excellent introduction to these issues in Daston and Galison 2010.

51  That there can be one such picture is, of course, a matter of furious debate. It will be discussed below.

52  Lynch 1998: 35.

53  Hacker 2003: 62.

54  Paul Snowdon discusses these accusations, in Snowdon 2009 (270).

55  See Rohlf 2014.

56  Yablo has a helpful overview of this and the realist position, in Yablo 2003.

57  Kolakowski 1968: 47–48 (quoted in Wiggins 2001: 149).

58  See particularly Wiggins 1980: Chapter 5 and 2001: Chapter 5.

59  See, for example, Robinson 2014 for an overview. See Wiggins 2001: 150.

> What is this substance out there that can be conceptualised in radically different ways, which can be seized upon in thought by the anti-essentialist, but can have radically different principles of existence and persistence ascribed to it? This is surely an entity with inconsistent properties.
>
> (Wiggins 2001: 148)

60  'Haecceity' is the medieval concept of *thisness* – that metaphysical quality that makes some object the specific object that it is. Whether it is as 'misbegotten' or as 'unlovely' as Wiggins says it is, it is certainly too confusing for me to usefully explicate here. See, for example, Wiggins 1980: 136–142.

61  Wiggins 2001: 146ff.

62  E.g. Wiggins 1980: 131, 2001: 141.

63  Yablo 2003: 1.

64  Wiggins 2001: 142.

65  Wiggins 2001: 150–151.

66  See specifically Wiggins 2001, Chapter 5.

67  Eddington 1958.

68  Wiggins 2001: 152.

69  Glock 2003: 19.

70  Wiggins 2001: 150–151.

71 See, for example, Wiggins's discussion in 2001: 140ff, see also Wiggins 2001: xii.
72 This comes from the sixth page of his introduction to *Twelve Essays* (2016) (yet to be paginated at Oxford University Press).
73 Wiggins 2001: 2.
74 Wiggins 2001: 3.
75 Wiggins 2001: 153.
76 See particularly Wiggins 2001: Chapter 3 (and the following section), and, for example, page 160.
77 Our expectations about objects are either met or they are not. See, for example, what befell 'phlogiston' (Wiggins 2001: 80).
78 This is an event that Wiggins sees to be most obvious in childhood development:

> The child who is learning to find for himself the persisting substances in the world, to think the thoughts that involve them and recognize the same ones again, grasps a skill and a subject matter at one and the same time.
>
> (Wiggins 2001: 2)

See also Wiggins 1997: 24: 'Infants are creatures who are *en route* by exploration, trial and error, by probation, by attunement, to the full human conceptual system.'
79 Wiggins 2001: 152.
80 See Wiggins 2001: xii.
81 It is considered in greater depth below. For now, however, it will do to have seen the general lines along which Wiggins aims to substantiate his descriptive approach. While his project is, as he admits, 'anthropocentric' (Wiggins 2001: 153), it is not too narrowly so, and does not fall into the snares of conceptualism (see Lowe 2003 and 2005 for discussion).
82 This 'conceptualist-realism' is one of Wiggins's key contributions to contemporary metaphysics (and one which is growing in influence – not least in the work of Helen Steward (see, for example, Steward 2012)).
83 Wiggins 2001: 3. See also responses to Fei Xu (Wiggins 1997), and 2012: 24, fn. 30: 'Think what damage would result from expunging the distinction between substance and process/narrative/event'.
84 This point is to be built upon in due course (and bolstered by his comments on Lewis §13–14 in Wiggins 2012).
85 Eddington 1928: ix–x.
86 Nagel 1986.
87 The notion of the 'absolute conception' of reality comes from Bernard Williams (and is found in Williams 1978).
88 Haack 1978.
89 Kolakowski 1968: 47–48.
90 The four dimensionalist's model of physics describes objects around us that are both spatially and *temporally* extended; as we have spatial parts (our hands and feet, etc.) we also have temporal parts (our infancy, our dotage and so on). David Lewis sees this model to provide clean and principled answers to puzzles of change, of how a single thing can have two intrinsic and apparently incompatible properties (e.g. sitting and standing). See, for example, Lewis 2002: 441. Cf. Wiggins 2001: 31. This is discussed in Chapter 7 in relation to the notion of *processes*.
91 Wiggins 2012: 12. See also his comments in his paper 'Identity, Individuation, and Substance':

> [I]t is not an option for philosophy to reject the four dimensional conception of the world urged upon us by some philosophers and metaphysicians of science. But in accepting it one is not committed to see things, people and organisms in perdurantist fashion as made up of instantaneous temporal parts.
>
> (2012: 135)

92  In this connection, see Wiggins's comments on Carnap in Chapter 1 of his latest collection (2016), fn. 27.
93  Strawson 1959: 9f, and Haack 1978.
94  Stebbing 1958: 51–52.
95  Thomasson 2007: 140.
96  There is another response available to Wiggins. If he holds that 'solid' is ambiguous, and functions differently in the scientific and pre-theoretical ideologies, he can assert that Table No. 1 and Table No. 2 are the same table spoken about in different ways with predicates taken in different senses. That Wiggins may take this line is suggested by his comments about Putnam in the introduction to his 2016 text (§5).
97  Wiggins 2001: 155–156 (my emphasis).
98  This is Button's term (found in Button 2013).
99  Wiggins 2016.
100  See, for example, Putnam 2004:

> [I]t is no accident that in everyday language we employ many different kinds of discourses, discourses subject to different standards and possessing different sorts of applications, with different logical and grammatical features – different 'language games' in Wittgenstein's sense – no accident because it is an illusion that there could be just one sort of language game which could be sufficient for the description of all of reality!
>
> (2004: 22)

101  Lynch 1998.
102  In opposing incompatibility, he finds a line of response to Donald Davidson's criticisms of 'conceptual schemes' (see Davidson 1974). For Davidson, the notion of a conceptual scheme was logically bankrupt – and in part, his argument for why this is the case, depends on the fact that different schemes purport to describe reality in different ways, without translatibility, and with potential for incompatibility (5). The frameworks described above exist at a greater level of generality – and Wiggins's position will be that there are no grounds for the accusation of incompatibility. He is not claiming that reality is relative to a scheme; objects can exist in one scheme just as much as another.
103  E.g. in Dupré 1993. See also, for example, Dupré 2010/2012: '[W]e should be pluralistic about how we divide the biological world into individuals: different purposes may dictate different ways of carving things up' (118). '[T]here will be no unequivocal way of … dividing reality' (126). For another take on this picture, see Rasmus Winther's discussion of 'partitioning frames' (Winther 2011).
104  Wiggins 2012: 21.
105  E.g. in Thomasson 2007.
106  McDaniel 2009.
107  Ingarden 1973.
108  As Wiggins himself states, however, one may carve a carcass in more than one way:

> Carving belongs with eating, one definite purpose, or with anatomy, another definite purpose. Maybe there is always a best way to carve for purposes of eating and a best way to crave in the cause of scientific anatomy. But as is evident from Plato's own example not all carvings and not all singlings out or classifyings … will serve the same theoretical or practical purpose.
>
> (This passage is taken from the draft of the introduction to his latest collection, *Twelve Essays*)

109  Quine 'On What There Is' in Quine 1980.
110  For replies to arguments based on parsimony, see Thomasson 2007: 151–175.
111  How do we examine in detail the persistence condition of a human being by turning to biology? This is the question asked in Chapters 6, 7 and 8.

## Bibliography

Ackrill, J.L. (1963) *Aristotle's Categories and De Interpretatione* (translation with notes) (Oxford: Clarendon Press).

Bagchi, K. (1980) *Descriptive Metaphysics and Phenomenology* (India: Prajna Press).

Beaney, M. (2015) 'Analysis', *The Stanford Encyclopedia of Philosophy*, E.N. Zalta (ed.) *(forthcoming)* available online at http://plato.stanford.edu/archives/spr2015/entries/analysis/.

Burtt, E.A. (1953) 'Descriptive Metaphysics', *Mind* LXXLL.

Button, T. (2013) *The Limits of Realism* (Oxford: Oxford University Press).

Byrne, P.H. (2001) 'Connective Analysis: Aristotle and Strawson', *British Journal for the History of Philosophy* 9(3): 405–423.

Chalmers, D., Manley, D. and Wasserman, R. (2009) (eds) *Metametaphysics: New Essays on the Foundations of Ontology* (Oxford: Oxford University Press).

Daston, L. and Galison, P. (2007) *Objectivity* (Cambridge, MA: MIT Press).

Davidson, D. (1974) 'On the Very Idea of a Conceptual Scheme', reprinted in *Inquiries into Truth and Interpretation* (Oxford: Oxford University Press, 1984).

Dupré, J. (1993) *The Disorder of Things: Metaphysical Foundations of the Disunity of Science* (Cambridge, MA: Harvard University Press).

Dupré, J. (2010/2012) 'The Polygenomic Organism', in J. Dupré *Processes of Life* (Oxford: Oxford University Press).

Eddington, A. (1958) *The Philosophy of Physical Science* (Michigan: Ann Arbour Paperbacks).

Eddington, A.S. (1928) *The Nature of the Physical World* (New York: Macmillan).

Glock, H.-J. (2003) 'Strawson and Analytic Kantianism', in H.-J. Glock (ed.) *Strawson and Kant* (Oxford: Oxford University Press).

Grundmann, T. and Misselhorn, C. (2003) 'Transcendental Arguments and Realism', in H.-J. Glock (ed.) *Strawson and Kant* (Oxford: Oxford University Press).

Haack, S. (1978) 'Descriptive and Revisionary Metaphysics', *Philosophical Studies* 35: 361–371.

Hacker, P. (2003) 'On Strawson's Rehabilitation of Metaphysics', in H.-J. Glock (ed.) *Strawson and Kant* (Oxford: Oxford University Press).

Hacking, I. (1968) 'A Language without Particulars', *Mind* LXXVII.

Haslanger, S. (2012) *Resisting Reality: Social Construction and Social Critique* (Oxford: Oxford University Press).

Ingarden, R. (1973) *The Literary Work of Art*, George G. Grabowicz (trans.) (Evanston, Illinois: Northwestern University Press).

Joll, N. (2010) 'Contemporary Metaphilosophy', *Internet Encyclopedia of Philosophy*, available online at www.iep.utm.edu/con-meta/.

Kolakowski, L. (1968) *Towards a Marxist Humanism* (New York: Gove Press).

Körner, S. (1966) 'Transcendental Trends in Recent Philosophy', *The Journal of Philosophy*.

Levesque-Lopman, L. (1983) 'Decision and Experience: A Phenomenological Analysis of Pregnancy and Childbirth', *Human Studies* 6(1): 247–277.

Lewis, D. (2002) 'Tensing the Copula', *Mind* 111: 1–14.

Lowe, E.J. (2003) 'Review of *Sameness and Substance Renewed*', *Mind*, New Series, 112(October): 448.

Lowe, E.J. (2005) 'Is Conceptualist Realism a Stable Position?' *Philosophy and Phenomenological Research* LXXI(2): 456–461.

Lynch, M. (1998) *Truth in Context: An Essay on Pluralism and Objectivity* (Cambridge, MA: MIT Press).

Mach, E. (1905) *Erkenntnis und Irrtum: Skizzen zur Psychologie der Forschung* (Leipzig: J.A. Barth).

McDaniel, K. (2009) 'Ways of Being', in D. Chalmers, D. Manley and K. Wasserman (eds) *Metametaphysics: New Essays on the Foundations of Ontology* (Oxford: Oxford University Press).

Mei, Tsu-Lin (1961) 'Subject and Predicate: A Grammatical Preliminary', *Philosophical Review* LXX.

Nagel, T. (1986) *The View from Nowhere* (Oxford: Oxford University Press).

Putnam, H. (2004) *Ethics Without Ontology* (Cambridge, MA: Harvard University Press).

Quine, W.V.O. (1980) *From a Logical Point of View* (New York: Harper and Row).

Robinson, H. (2014) 'Substance', in E.N. Zalta (ed.) *The Stanford Encyclopedia of Philosophy*, *(forthcoming)* available online at http://plato.stanford.edu/archives/spr2014/entries/substance/

Rohlf, M. (2014) 'Immanuel Kant', in E.N. Zalta (ed.) *The Stanford Encyclopedia of Philosophy* available online at http://plato.stanford.edu/archives/sum2014/entries/kant/

Ryle, R. (1970) 'Autobiographical', in O.P. Wood and G. Pitcher (eds) *Ryle: A Collection of Critical Essays* (New York: Doubleday).

Russell, B. (1925) *Mysticism and Logic* (London: Longman's, Green & Co.).

Skinner, Q. (2002) *Visions of Politics: Vol. 1: Regarding Method* (Cambridge: Cambridge University Press).

Snowdon, P. (2009) 'On the Sortal Dependency of Individuation Thesis', in H. Dyke (ed.) *From Truth to Reality: New Essays in Logic and Metaphysics* (London: Routledge).

Stebbing, S. (1958) *Philosophy and the Physicists* (New York: Dover).

Steward, H. (2012) *A Metaphysics for Freedom* (Oxford: Oxford University Press).

Strawson, P.F. (1950) 'On Referring', *Mind*, New Series, 59(23): 320–344.

Strawson, P.F. (1959) *Individuals* (London: Methuen).

Strawson, P.F. (1966) *The Bounds of Sense: An Essay on Kant's Critique of Pure Reason* (London: Methuen).

Strawson, P.F. (1992) *Analysis and Metaphysics: An Introduction to Philosophy* (Oxford: Oxford University Press).

Thomasson, A. (1999) *Fiction and Metaphysics* (Cambridge: Cambridge University Press).

Thomasson, A. (2007) *Ordinary Objects* (Oxford: Oxford University Press).

Whitehead, A.N. (1919) *The Concept of Nature* (Cambridge: Cambridge University Press).

Wiggins, D. (1980) *Sameness and Substance* (Cambridge, MA: Harvard University Press).

Wiggins, D. (1997) 'Sortal Concepts: a Reply to Fei Xu', *Mind and Language* 12: 413–421.

Wiggins, D. (2001) *Sameness and Substance Renewed* (Cambridge: Cambridge University Press).

Wiggins, D. (2012) 'Identity, Individuation and Substance', *European Journal of Philosophy* 20(1): 1–25.

Wiggins, D. (2016) *Twelve Essays* (Oxford: Oxford University Press).

Williams, B. (1978) *Descartes: The Project of Pure Inquiry* (London: Pelican).

Winther, R.G. (2011) 'Part-whole Science', *Synthese* 178: 397–427.

Wynn, F. (2002) 'The Early Relationship of Mother and Pre-infant: Merleau-Ponty and Pregnancy', *Nursing Philosophy* 3(1): 4–14.

Yablo, S. (2003) 'Tables Shmables: Review of David Wiggins, *Sameness and Substance Renewed*', in the *Times Literary Supplement* (July).

# 2 D

The methodological background is in place. Wiggins's project falls within the descriptivist tradition where metaphysical inquiry involves analysis of our pre-theoretical framework. It is against this backdrop that Wiggins constructs his thesis **D** – the subject of the present chapter.

Thesis **D** is a thesis about the fundamental metaphysical relation of *identity*. What is *identity* exactly? What does it mean for some *x* to be *identical* to some *y*? The relation, represented in mathematics with the '=' symbol may seem of somewhat limited interest, but one of the charms of Wiggins's account is that it shows the extent to which the notion transfuses and supports our everyday lives. The thought that entities are *identical* cannot be done without. We would lose track of ourselves, of those we love, of our very position in the world, if we could not say that they do, or do not, stay 'the same' through time.

Yet **D** does not only involve claims about the logic of that relation; it encompasses, equally, a thesis about *individuation*. Individuation is, as Wiggins puts it, 'something done by a thinker'.[1] It is a cognitive, or psychological act or process, in which a thinker picks out objects in her environment.[2] When we pick out a friend and follow their progress through a crowd, we are performing an individuative act: we are singling something out, in such a way that it can be traced and re-identified. So **D** is not an isolated thesis about the mysterious connection we invoke when we write '='; it comprises a thesis about our everyday experience of the world. This chapter aims to emphasize how deeply interconnected he sees these theses to be, and to demonstrate their inter-dependence.

In rough outline – with specialist vocabulary to be refined in due course – the exposition will proceed as follows: it will be shown how Wiggins develops a skeletal account of identity from the basic properties that are contained in our pre-theoretical grasp of it. But while such an account provides *necessary* conditions for that relation, it does not furnish a *sufficient* condition. Wiggins proposes to fill this lacuna by turning to our individuative practices. These practices depend crucially upon the *kind* or *sort* of thing that is to be traced through space and time and the sort or kind directs us to the individual *principles of activity* that substances exemplify. It is by these that we assess persistence. The so-called 'D-principles' set out the consequential requirements upon a substance sortal concept.[3]

## 1  The bare bones of identity

In the descriptivist spirit, Wiggins turns first to our pre-theoretical idea of identity.[4] It is this primitive concept, so fundamental to our conceptual framework, which he aims to elucidate.[5] *Identity*, he claims, is one of the core concepts that all human inquirers are supposed to possess (another demonstration of his commitment to the thesis of 'conceptual invariance', announced in Chapter 1):

> The notion of sameness or identity that we are to elucidate is ... a notion as primitive as predication.... No reduction of the identity relation has ever succeeded.[6]

We are not talking about the notion of a *character* (or something like it), which is in certain circumstances used synonymously with 'identity' (in cases of 'assumed identities' or 'fake identities'). 'Identity cards' used at border controls are not the focus of this study. We are talking about a *relation*. It is a relation that holds between Aristotle and the author of *De Anima*, the relation that holds between the individual (you) who started reading this sentence, and the individual (you again) who has just now finished it. It is the relation that holds between something and itself.

This idea of some *x* *being the same as* some *y* is basic to our everyday navigation of the world. It gives us surety in our movements. Think of the routine task of finding food items in a supermarket; walking up and down the aisles, you trust that the signs for baked goods *stay the same* even when you lose sight of them. If we thought these way-markers did not *stay the same*, we would move around with much less confidence than we do. The same is true for way-markers of every kind, of places and thoughts and people.

Wiggins's aim is to elucidate the idea of *sameness*. He starts his inquiry with what he sees to be its self-evident properties. The first is the so-called *reflexivity* of identity – 'the obvious truth where everybody begins'.[7] This holds that: For all $x$, $x$ is the same thing as $x$. You will not get far denying that $x$ is $x$, that Aristotle is Aristotle or that the number 42 is the number 42.

The second self-evident property is Leibniz's principle of the *indiscernibility of identicals* – 'akin to the rock of ages'.[8] This requires that if $x$ is the same thing as $y$, then $x$ and $y$ have all the same properties (or, more formally, 'if $x$ is identical with $y$, then $x$ is f if and only if $y$ is ... f'). This, too, is hard to resist. Let us say, as was the case, that George Eliot is Mary Ann Evans. George Eliot is the author of *Middlemarch*; the same holds true for Evans. Indeed, there is no property that Eliot has that Evans lacks: George Eliot was a woman who assumed a man's name in order to publish her manuscript; so was Mary Ann Evans. George Eliot was born in 1819; so was Mary Ann Evans. Since George Eliot *is* Mary Ann Evans, there is no property that the latter has that the former does not also have, nor vice versa.

Wiggins uses these two aspects of our primitive notion as the foundation for further analysis.[9] From the combination of the two original principles he first of all reaches the *transitivity* and *symmetry* of identity. *Transitivity* holds that if *x* is *y*, and *y* is *z*, then *x* is *z*. How so? Well, if *x* is *y*, and *y* is *z*, then any property *x* has *y* has – and since *y* is *z*, *x* too is *z*. This is a bit schematic, but think again of Eliot and Evans. Eliot is the author of *Middlemarch*; Evans is Eliot; thus Eliot is the author of *Middlemarch*. The property of *symmetry* similarly develops from the first two principles, and holds that if *x* is *y*, then *y* is *x*. (Eliot is Evans, so Evans has all of the properties that Eliot has. One of these properties is that, being Eliot, Evans is Eliot.)

Following on from these properties are the *necessity*, the *absolute determinacy* and the *permanence* of identity. *Necessity* holds that 'If *x* is the same thing as *y*, then *x* is necessarily the same thing as *y*'.[10] *Determinacy* holds that 'If *x* is the same thing as *y*, then *x* is absolutely and determinately the same thing as *y*'[11] (which is to say there is no vagueness when it comes to identity – though there might well be vagueness in our appreciation of the fact that *x* is *y* (I might not be *sure* that Eliot is Evans)). *Permanence*, finally, states that 'If *x* is the same thing as *y* then *x* is always the same thing as *y*'.[12] These are the bare bones of the identity relation.

## 2 The 'epistemology of the relation'

There is much more that can be said about the formal properties of identity but this is not the place to say it.[13] My aim is only to show how Wiggins unpacks this logical relation.[14] His is a worthwhile project – but a troubled one too. The problem is that the formal properties provide no sufficient condition.[15] (Note that it is the sufficient condition that the *personal identity* theorist is interested in, needing a workable condition on which to ground her judgements of identity.)

This 'deficit' Wiggins writes, 'is often overlooked because it will appear that we can safely add to Leibniz's Law the Leibnizian converse, namely the Identity of Indiscernibles'.[16] That principle states that if *x* and *y* have all their properties in common, then *x* is *y*. This principle is taken to flow from the logic of identity and to stand as a sufficient condition for it. However, as Wiggins rightly emphasizes, since the complete community of properties flows *from* identity, it cannot therefore, be taken to rule for or against it.[17] The indiscernibility of *x* and *y* could only be verified

> at the conclusion of explorations and labours conducted on the basis of a workable sufficient condition based in some thought other than the Identity of Indiscernibles.[18]

Being the mortal beings we are, we cannot establish identity between objects by examining *all* of their properties, making sure they correspond.[19] Before knowing that George Eliot is Mary Ann Evans, I would not be able to find out for certain whether they possessed all of the same properties, no matter how closely I studied them. Our use of the identity concept must appeal to *another* principle, one that can be applied by an enquirer in adjudicating persistence questions.[20]

And here we begin to see the effects of the connective analysis described in Chapter 1. Given that the identity of indiscernibles will not bear the weight commonly assigned to it, Wiggins directs us to an alternative:

> [T]he metaphysics of identity has no alternative but to reconstruct the thoughts that organize the epistemology of the relation and to reconstruct what thinkers actually do when they single out an object in experience, at once observing the thing's behaviour, speculating what it does when out of view and searching for distinguishing marks (if any) by which this one may be distinguished from other members of its kind and (however fallibly) reidentified as one and the same.[21]

We must turn, that is, to our *individuative* practices and the 'epistemology of the relation'. Individuation is the cognitive act of a thinker. More precisely, for Wiggins, it is the act of *singling out* or *picking out* an object in a thinker's environment, and picking them out as objects with spatio-temporal limits. That is, as *continuants*.[22] In the introduction to *S&SR*, we find the following explanatory passage:

> The *Oxford English Dictionary* defines 'individuate' in terms of 'single out' or 'pick out', and this definition is well suited to the purposes of this book.... To single *x* out is to isolate *x* in experience; to determine or fix upon *x* in particular by drawing its spatio-temporal boundaries and distinguishing it in its environment from other things of like and unlike kinds (at this, that and the other times during its life history); hence to articulate or segment reality in such a way as to discover *x* there.[23]

When you looked around and found this book, and fixed on it as something to read, you performed an individuative act. You singled it out in the experiential stream and you kept track of it. Disappointed or cheered by its contents you have continued to interact with it; understanding the kind of thing that it is, you know roughly what it takes to destroy it (e.g. burning), what it can undergo (e.g. mild water damage) and so on. Such individuative procedures as this are the means by which we navigate the world.[24]

It is this process that Wiggins wants to understand more clearly. And subjecting it to closer scrutiny, he finds a question constantly recurring. When directing attention to objects in our environment and assessing their coincidence, we find ourselves asking questions like: *what is it?*[25] *What* kind *of thing is* x*?*[26] These questions, he suggests, give us clues as to what organizes our efforts when we identify and re-identify objects, when we *individuate*. He sees us constantly called upon to ask what the objects under investigation *are*. Thus, this is how he starts his investigations in *S&SR*:

> [I]t seems certain ... that, for each thing that satisfies a predicate such as 'moves', 'runs', or 'white', there must exist some known or unknown,

named or nameable, kind to which the item belongs and by reference to which the 'what is it' question *could* be answered.[27]

Wiggins takes the recurrence of this 'what is it?' question to suggest that matters of sameness revolve in some fashion, on the sorts, or 'sortal predicates', under which the objects fall.[28] Sortal terms take numerical modifiers – we use them to count items (e.g. 'cat' rather than 'air'). We pick individuals out 'under' sortals. Sortal *concepts* are the concepts associated with that kind. Sortal *predicates* are those expressions that stand for these concepts.[29] For Wiggins, a sortal predicate is an expression that gives a criterion for counting items of a kind, but *also* gives criteria of identity for members of that extension. The full import of these, admittedly somewhat technical, specifications will appear in due course.

Significantly, Wiggins also separates out two different types of sortal. There are *phased sortals*, which an individual may fall under at one time and not another – e.g. you may, temporarily be a student and then stop being one (when you no longer pay your tuition fees). The same is true for being a toddler, or a soldier, or a nurse, etc.[30] There are, however, *substance sortals* too, such as *human being*. If an individual falls under a substance sortal, it cannot cease to do so and continue to exist. They apply to an individual *x* at every moment through *x*'s existence.[31] Contrast a phased sortal. Phased sortals are typically 'restrictions' of more general sortal terms.[32] Substance sortals are what Wiggins says we have to be interested in when we are examining questions of identity.

Let us return to the main thread. Examining our individuative procedures, we find the 'what is it?' question constantly recurring. In judging identity, we typically try to understand what the candidates for identity *are*. Why is this 'sortal' approach so central?

> Suppose I ask: Is *a*, the man sitting on the left at the back of the restaurant, the same person as *b*, the boy who won the drawing prize at the school I was still a pupil at early in the year 1951? To answer this sort of question is surprisingly straightforward in practice, however intricate a business it would be to spell out the full justification of the method we employ. *Roughly, though, what organizes our actual method is the idea of a particular kind of continuous path in space and time the man would have had to have followed in order to end up here in the restaurant.... Once we have dispelled any doubt whether there is a path in space and time along which that schoolboy might have been traced and we have concluded that the human being who was that schoolboy coincides with the person/human being at the back of the restaurant, this identity is settled.*[33]

It is an anodyne example, and purposefully so. This is an everyday scenario, and one that we can deal with without recourse to complex metaphysical theories. We have the resources to cope with this identity question; our method is organized by 'the idea of a particular kind of continuous path in space and time the man would have had to have followed'. The essential point is that

[t]he continuity or coincidence in question here is that which brought into consideration *by what it is to be a human being*.[34]

'In practice', our answers to identity questions are organized by the idea of *spatio-temporal continuity*. Spatio-temporal continuity is what matters, not the idea of the total agreement of properties and relations.[35] And, crucially, the thought underpinning our practical reasoning is that spatio-temporal continuity is not, as we find in experience, *bare* continuity.[36] It is the specific kind of continuity that relates to the sort of thing that the objects are. When we draw spatio-temporal boundaries, we look to what a thing *is* ('Bare continuity supplies no principle, no rhyme or reason'[37]).

From all this it follows that we must be ready to ask after the kinds of changes a thing can suffer, how exactly it continues through space and time. Is this ice cube the same as the ice cube I took from the freezer an hour ago? If it is a hot day and there is no air conditioning, chances are it is not (the original would have melted); if I am in Alaska, in the depths of winter and without a radiator, it is quite possibly the same ice cube. I make these judgments on the basis of my understanding of what it takes for such things to persist. We find a general statement of this position at the start of Wiggins's 2012 paper 'Identity, Individuation and Substance':

> [I contend] that the key to this problem rests at the level of metaphysics and epistemology alike with a *sortalist* position. Sortalism is the position which insists that, if the question is whether *a* and *b* are the same, it has to be asked *what are they?* Any sufficiently specific answer to that question will bring with it a principle of activity or functioning and a mode of behaviour characteristic of some particular kind of thing by reference to which questions of persistence or non-persistence through change can be adjudicated.[38]

## 3  Principles of individuation

Thus Wiggins holds that each such (substance) sortal encapsulates a particular 'mode of being' or 'nature' (in the case of living things) or a 'function' (in the case of artefacts), on the basis of which we can adjudicate questions about their persistence.[39] The distinction between natural things and artefacts is an important one, to which we will return in Chapter 3 – but for the time being let us focus on the former. For *natural things*, the nature is the *mode of being* for things of that sort:

> It is the principle of activity of a kind whose members share and possess in themselves a distinctive source of development and change.[40]

Take the sortal *cat*, for example. Wiggins will say that there is a distinctive 'mode of being' for cats, a Leibnizian 'law' of activity, to which members of that kind are subject.[41] And this 'law' or 'principle' captures, ultimately, what it is to be a cat, from typical behaviour to determinate patterns of growth and development,[42] eating habits, metabolic processes, social (or, more usually, anti-social) tendencies and so on.[43] In each particular cat these standard capacities are determined in more specific ways, from its particular gait to the way its jumps, and purrs, and naps. And this specialized and refined actualization of the characteristic activity for *felis catus* is what ordinary thinkers (like us) have in mind (however imperfectly[44]) when considering such creatures' spatio-temporal continuity. It is the continuity of this specific actualization that counts for spatio-temporal continuity.

Such, then, is the special effectiveness of the 'what is it?' question:

> [I]n the case of continuants it refers us back to our constantly exercised idea of the persistence and life-span of an entity.[45]

This analysis allows Wiggins to add flesh to the logical bones bequeathed to us by Leibniz. It gives us a workable sufficient condition for identity (formalized, below, in Wiggins's **D**(iii)). And, as the rudimentary practices guide us towards a sufficient condition, those same everyday individuative practices are shored up against the formal metaphysics of identity. Drawing out that which underpins our pre-theoretical conceptions, Wiggins constructs a theory that satisfies the exigent logical requirements of that relation.[46] To this end, in Chapters 2 and 3 of the *Sameness and Substance* books, Wiggins sets out his **D**-principles:

> D-principles transcribe or transpose the formal properties of identity into universal norms to which all singling things out, all acts of recognizing the same again, all reconstructing the histories of things not continuously observed, and all judgment about identity and difference have, on pain of our losing hold altogether of reference and identity, to make themselves singly and collectively answerable.[47]

These **D**-principles set the requirements that must be met if our everyday individuative acts are to correspond to the formal properties of identity. A brief overview of these principles will be sufficient here (to be refined below, where necessary). Principles **D**(i)[48] and **D**(ii) are absorbed in **D**(iii) as follows:

> D(iii): *a* is identical with *b* if and only if there is some concept f such that (1) f is a substance-concept under which an object that belongs to f can be singled out, traced and distinguished from other f entities and distinguished from other entities; (2) *a* coincides under f with *b*; (3) [coincides under f] stands for a congruence relation: i.e. all pairs $\langle x,y \rangle$ that are members of the relation satisfy the Leibnizian schema $\Phi x$ if and only if $\Phi y$.[49]

The subsequent principles – **D**(iv)–**D**(x) – formulate exactly what it is to be a substance sortal (or rather, the associated substance concept) which can stand as an effective foundation for individuative acts. We need not go into all these in depth, but we will consider **D**(v) and **D**(vi), which play significant roles in the following sections. First though a little more on **D**(iii).

As it stands, the sufficient condition formalized in **D**(iii) marks the distinction between phased and substance sortals. But as Wiggins notes, there are apparent substance sortals which will not do the work we assign to them in this context:

> There are countless predicates in English that have the appearance of sortal predicates but are purely generic (*animal, machine, artefact*) or are pure determinables for sortal determination (*space-occupier, entity, substance*).[50]

These apply to *x* at every moment of *x*'s existence, but they are not suitable for individuation. They do not fully answer the 'what is it?' question because they do not specify what matters turn on in regard to persistence – i.e. they do not allow us to trace spatio-temporal continuity.[51] Consequently, then we find **D**(v) which sets out the all-important requirement for a principle of activity/functioning/operation:

> D(v): f is a substance-concept only if f determines either a principle of *activity*, a principle of *functioning* or a principle of *operation* for members of its extension.[52]

These principles, as mentioned, are 'principles of individuation' – principles 'by which entities of [that] particular kind may be traced or kept track of and re-identified as the same'.[53] If we are trying to individuate artefacts (chairs, say, or tables), we turn to their principle of *functioning*. For natural organs (hearts, livers), we turn to their principle of *operation*. And when we are trying to pick out and trace natural substances, like cats, or rhododendrons, or human animals, we turn – as we have seen – to their distinctive principle of *activity*.

Also worthy of mention here is **D**(vi). This principle – relevant to the discussions of *fission* that appear in Chapter 9 – states that f is a substance concept only if it determines a notion of coincidence (or in the case of changeable substances, a notion of continuity[54]) that is fully *transitive*:

> D(vi): If f is a substance concept for *a*, then *coincidence under f* is fully determinate enough to exclude this situation: *a* is traced under f and counts as coinciding under f with *b*, *a* is traced under f and counts as coinciding under f with *c*, yet *b* does not coincide under f with *c*.[55]

**D**(vi) emphasizes another distinction between merely apparent and genuine substance sortals. It distinguishes between sortals that cover substances – changeable, persisting substances of the kind we find around us, i.e. *particulars* – and sortals that apply to clones, or varieties, or strains, i.e. *universals*. (The notion of a 'universal' is elaborated in Chapter 9.)[56] Consider, for example, the amoeba that splits. Neither of the products can be the same amoeba as the 'parent'. They exemplify the same strain or clone, but no more.[57] As we will see (again, in Chapter 9), this analysis feeds directly into Wiggins's recent assessment of brain transplantation. We return to these matters in the next chapter, which focuses on the metaphysical distinction between natural things and *artefacts*.

We might also mention – for the sake of completeness and as a point of interest – the final **D**-principle: **D**(x), or the 'Only *a* and *b*' rule:

> D(x): ...the identity of *a* with *a*, of *b* with *b*, and of *a* with *b*, once we are clear which things *a* and *b* are, ought to be a matter strictly between *a* and *b* themselves.[58]

This is to say that if *a* is really *b* then their identity will hold irrespective of anything else. The existence of another thing *c* will not suddenly change things. This demand becomes critical in relation to the case of Theseus' ship (discussed in Chapter 3). If the ship that stands in the port today is truly the same as the ship which Theseus piloted, even though its parts have been utterly overhauled, then this identity must not be disturbed by the appearance of another ship, of exactly the same make, built from the discarded parts of the original. And yet, if we find this second pretender convincing, then things *will* be disturbed – since they cannot *both* be identical with the original. In this way we are given to question the stability of the original identity claim.

**D**(x) is nevertheless much controverted.[59] Richard Gale, in his 1984 paper 'Wiggins's Thesis D(x)',[60] attempts to discredit it – but seems to do so on an broadly epistemic understanding of the claim ('the most natural interpretation [of it] is that observations of objects other than *a* and *b* cannot either verify or falsify (or confirm or disconfirm) that *a* [is identical] to *b*'[61]). Doing so, Gale misses the point, as Wiggins makes clear when he writes that 'the identity of *x* and *y* may of course be discovered in all sorts of ways that involve reference to other things, but constitutively the identity of *x* and *y* involves only *x* and *y*'.[62] Nonetheless, for the sake of Gale and other critics, Wiggins has subsequently reformulated **D**(x) into what he calls an 'Adequacy requirement',[63] which he sees as an indispensible improvement ('whatever grounds the identity of *x* with *y* must ipso facto and, and by that same token, ground the indiscernibility of *x* and *y*'[64]).

*Sameness* is a pre-theoretical notion; we use it ceaselessly, and we do so effectively. Our methods for judging whether or not some earlier thing is *the same as* some later thing are reliable, and being so, can be used – as Wiggins uses them – to enrich a formal rendering of the *identity* relation. But it is important to be aware of the rules that organise our usage of 'same'; why does it work, when it works? What laws must our usage abide by, if we are to use it coherently? For Wiggins, the principles that undergird individuation need to be set out in a way that is logically hygienic: this is the aim of the **D**-principles. Individuation nourishes identity and identity, in turn, nourishes individuation.

*   *   *

Few philosophical theories have been so carefully cut nor finely polished as David Wiggins's thesis **D**. It is multi-faceted; it encapsulates claims about the logic of identity and its formal properties, while simultaneously capturing thoughts about our epistemic practices. Its beauty is in showing how these ideas are different faces of the same stone; we cannot understand *identity* without simultaneously understanding *individuation*. This sketch will bear some refinement – and this is pursued in Chapter 3 (particularly in relation to the concept of *substance*, which also falls within the ambit of **D**). Now, however, we will do well to consider some objections and some novel interpretations of Wiggins's thesis. The first is a general concern with the kind of sortalism he is offering, the second is a more precise worry from Paul Snowdon, a sympathetic but not uncritical commentator on Wiggins's work.

## 4  Sortal identity and relative identity

There is a problem with the kind of sortalism Wiggins describes: it is often taken to lead towards a commitment to *relative* identity. In what way? It is clearly the case that we can pick out the same object as different sorts of things – e.g. *man, soldier, greybeard, Greek, philosopher, official of the state, the chairman for day d, 406 B.C.*, etc. … And surely, we may say, it makes all the difference to claims of identity *which* of these concepts one subsumes the something under.[65] An individual *a* might be the same *man* as individual *b*, but not the same *official of the state* (if, say, *a* lost his title). So this relativism ushers us inexorably towards the kind of anti-realist conceptualism described in the first chapter,[66] where identity is seen to be relative to us and not answerable to the world.

Wiggins intends his treatment of *sameness* to accommodate both the insights of realism and those of conceptualism. The sortalism that emerges from our everyday practices must be answerable to the 'metaphysics of identity', and prime among the formal properties of that relation is Leibniz's principle of the Indiscernibility of Identicals. The *relativity* of identity – as advanced by Peter Geach and Harold Noonan[67] – marks the contravention of this principle.[68] Consequently, the focus of the first chapter of *S&SR* is to reconcile Wiggins's sortalism with strict Leibnizian identity. It will be useful to outline his explanations of apparently relativistic identity claims – specifically his appeal to phased sortals, and the 'is' of constitution.

Wiggins holds that some sortals are, in a certain sense, more fundamental than others. In his early work he speaks of 'ultimate' sortals,[69] but this term has since been dropped. The kinds of sortal in question are *phased* sortals, under which an individual may fall temporarily, and *substance* sortals, which that individual cannot fail to fall under and continue to exist. The connotation of 'substance' in Wiggins's texts is drawn out in the next chapter; the focus here is on how this distinction gives him the resources to answer certain counter-examples to absolute Leibnizian identity. A sortalist may say that an individual *x* is the same *human being* as *y*, but not the same *nurse* (he was once a nurse but has since retired): the identity of *a* and *b* thus seems to turn on which sortal we subsume them under – so we are drawn towards relativism. But this conclusion is only reached if, says Wiggins, we ignore the *sorts* of sortals at play:

[One must] distinguish between sortal concepts that present-tensedly apply to an individual *x* at every moment throughout *x*'s existence, e.g. *human being*, and those that do not, e.g. *boy*, or *cabinet minister*. It is the former ... that give us the privileged and ... the most fundamental kind of answer to the question 'what is *x*?' It is the latter, one might call them phased-sortals, which, if we are not careful about tenses, give a false impression that *a* can be the same f as *b* but the same g as *b*.[70,71]

Thus, relativism can be read – in some cases at least – as emerging from confusions of phased sortals with the substance sortals they restrict. You might say that Tony Blair is the same *man* that he was in the 1990s while denying that he is the same *prime minister*, but this is a function of the phased nature of the sortal *prime minister*; Blair is no longer prime minister. This does away with some of the problematic cases. What about the others?

Wiggins has another recourse by which to discredit apparent relativity, namely the so-called constitutive 'is'. Consider another oft-cited story that appears to support the relativist's thesis ('**R**' as it is called in the literature): having broken a jug – reduced it, by accident or malice, to a heap of fragments – you (conscientiously) cement the pieces back together ... but (mischievously) into a coffee-pot this time, and not a jug. The relativist will claim that here we have a case where 'the jug is the coffee pot' is true with the covering concept *same collection of material bits*, but false with the covering concept *same utensil*.[72] So Leibnizian identity is again called into question.

Wiggins's response to this kind of example appeared first in his examination of material coincidence in his paper 'On Being in the Same Place at the Same Time'.[73] There, he outlines a 'metaphysics of constitution' that distinguishes between two uses of the word 'is'. 'Is' can represent an identity relation (e.g. George Eliot *is* Mary Ann Evans) – and this is how 'is' is usually taken. So when we state that the jug *is* the heap of fragments, and the heap of fragments *is* the coffee pot, we seem to be claiming identity (which must then appear relative since the coffee-pot and the jug cannot have the same properties (being different utensils)). Yet Wiggins claims that 'is' can also represent the much weaker 'constitution' relation. Notably, whereas identity is symmetric and reflexive (as described above) the constitution relation is asymmetric and irreflexive. The heap constitutes the coffee-pot and the jug, but neither the coffee-pot nor the jug constitute the heap.[74] Wiggins's re-description of the story then, is as follows: *x* and *y* are not identical under the sortal *utensil* (f) (since one is a coffee-pot and one a jug). But neither are they identical under the sortal *collection of fragments* (g), since neither *x* nor *y* is identical with the collection of fragments – they are merely *constituted* by that very collection.

In saying this, Wiggins invites a further objection. According to the proposed analysis, we find two material things (the lump and the jug) inhabiting the same spatio-temporal region. This appears to stand in defiance of common-sense:

It is a truism frequently called in evidence and confidently relied upon in philosophy that two things cannot be in the same place at the same time.[75]

This is the truism with which Wiggins's begins 'On Being in the Same Place at the Same Time'. Therein, he defends the independence of entities held together by the constitution relation (he describes, for example, a tree and the aggregate of cellulose molecules that make it up, and the mismatch of their persistence conditions), before issuing an amendment to the supposed truism. The amendment runs as follows:

> No two things *of the same kind* (that is, no two things which satisfy the same sortal or substance concept) can occupy exactly the same volume at exactly the same time.[76]

It is an important qualification, and one which the recent history of Anglophone philosophy has found to be tremendously persuasive. We need not rehearse his arguments here but it is worth noting that the kind of material coincidence this reformulation allows is now well established in the personal identity literature. It is the foundation for a variety of classic puzzles – like those presented by Eric Olson and Rory Madden[77] – of coincident thinkers. It stands as a central strut in the kind of 'constitutionalism picture' described by Lynne Rudder Baker.[78] You and I cannot stand in the same place at the same time, no matter how hard we try; but you and a certain aggregate of flesh and bones and blood are certainly coincident.

Another point worth noting is that this kind of 'material coincidence' plays a specific role in Wiggins's verdant metaphysical realism. He countenances a whole gamut of entities and this can lead us to issues with *co-habitation*. Sat where I am currently sat it may be claimed that there are a number of different beings: a human being, an aggregative entity, a processual entity (such as the working of the krebs cycle and the processes that depend upon it). These things are surely material things but they are not the same *kind* of material thing. They do not have the same metaphysical character.

Once again, let us allow ourselves to be persuaded by Wiggins's explanation. Let us agree with him that identity is identity under a sortal concept, and agree that this does not, in the end, commit him to a relativization thesis. For the time being, at least, let us say that any appearance of relativity may be explained away in a manner that conforms to the requirements of Leibnizian identity.[79] It is time to move onto criticisms offered by Paul Snowdon.

## 5   A metaphysical or a psychological thesis?

> At the centre of this web [of metaphysical ideas] is a thesis which Wiggins calls the thesis of the sortal dependency of individuation (2001: 5), though at times he seems prepared to call it the sortal dependency of identity (2001: 23), and which he usually refers to as D.[80]

This quotation comes from Snowdon's paper 'On the Sortal Dependency of Individuation Thesis', in which he broaches this apparent equivocation which he finds at the heart of *S&SR*. Snowdon is an insightful commentator, and while in his own work he pursues an animalist agenda, his analyses of Wiggins's texts have always been sympathetic and fruitful. (As will appear in Chapters 4 and 9, his commentaries have encouraged Wiggins to refine his position on *persons*.) Snowdon discerns two main theses here: a theory about individuation and a theory about identity. He assesses each in turn. He presses the question whether Wiggins is presenting a *metaphysical* theory (relating to identity) or a *psychological* one (relating to individuation).[81] His thought is that there are claims about both, and that while the identity theory is strong, the – psychological – theory of individuation is deeply problematic.[82]

This reading clearly diverges from the interpretation I have given above. Snowdon takes the apparent equivocation of 'sortal dependency of individuation' and 'sortal dependency of identity' to be a confusion in Wiggins's work. Moreover, he advises that we should preserve 'the distinction between the two sorts of claims and abandon the psychological claims'.[83] I want to suggest however that the 'sortal dependency of individuation' and the 'sortal dependency of identity' are not, for Wiggins, separable from each other.[84]

I began by remarking that Wiggins is working firmly within the descriptivist tradition. He sees our conceptual framework as the litmus of metaphysical inquiry. A descriptivist will need to be persuaded of the plausibility of Snowdon's partitioning of the metaphysical from the psychological. Thesis **D** will not be exclusively 'metaphysical' insofar as it is attuned to human cognition and human conceptions.[85] Moreover, a psychological thesis can stretch beyond the purely cognitive. As Wiggins puts it in Quinean parlance, he rejects the 'sharp division of questions of ontology from questions of ideology'.[86]

This response takes on a more precise form when we turn to the nuances of Wiggins's descriptivism. Throughout his work he emphasizes how complex our pre-theoretical framework actually is, and how inextricably bound up the concept of identity is with our individuative practices, and how, in order to get a tangible grasp of identity, we need also to grasp how 'continuants' are picked out and traced through time.

> These things belong together – or the grasp of each requires the grasp of the other.[87]
>
> These practices are intertwined with one another .... [T]he basic forms and devices have to be learned together. Just as the keystone of an arch and the adjoining bricks can be placed together, but only if somehow they are placed simultaneously or they are put into position with the help of a temporary external support, so each primitive device is learned *simultaneously and in reciprocity* with each of the others.[88]

Indeed, this is one of the most attractive aspects of Wiggins's account; he recognizes the complexity and interconnectedness of our conceptual practices, and how subtly interlaced, and mutually dependent our thoughts actually are. Here again we see Strawson's influence. As Strawson set it out, 'connective analysis' elucidates concepts by drawing out their connections and studies the way in which they imply, presuppose, and sometimes exclude one another.[89] The equivocation that Snowdon identifies in *S&SR* (and its predecessors) is not a confusion. It is a central element of Wiggins's project. The two elements of **D** are intended to develop reciprocally and synchronically. Thus, when Snowdon writes 'It seems clear that the minimal metaphysical principle could be true without the psychological one being true',[90] he misses the point. They are, for Wiggins, two sides of the same coin.

Let us look more closely. Consider, Snowdon's starting concern with Wiggins's project:

> When Wiggins states D for the first time in a relatively full way, in Chapter 2 of *Sameness and Substance Renewed*, it is formulated as follows:
>
> $a=b$ if and only if there exists a sortal concept f such that
>
> 1  *a* and *b* fall under f;
> 2  to say that *x* falls under f or that *x* is an f is to say what *x* is (in the sense Aristotle isolated);
> 3  *a* is the same f as *b*, that is coincides with *b* under f in the manner of coincidence required for members of f, hence congruently.
>
> (Wiggins 2001: 56)

> Taking this formulation of D as the guide to its general character as a claim, it is natural to comment on its significance as follows. D advances a claim about the necessary and sufficient conditions for the truth of *any* identity proposition or claim. As it stands, D is, strictly speaking, an unmodalized biconditional, but we can, surely, take it that Wiggins is really advancing it as a necessary truth about identity. Now one thing that stands out about D as formulated is that it says nothing about our *procedures* for establishing identities, nor does it say anything about our ability to *grasp or understand* identity propositions. Hence, if 'individuation' stands for something cognitive or psychological that we *do*, or maybe, *achieve*, as I claim that it in fact does, then it seems that D actually says nothing about *individuation* at all. So there is some inclination to register at least mild surprise about the name Wiggins assigns to it. Rather, I think, D says something about the conditions for the obtaining of an identity and so qualifies as what I would be inclined to call a '*metaphysical thesis*' about identity.[91]

Snowdon is right that this formulation stands as a claim about the metaphysical relation of identity. And while he is tentative in describing it as the 'sortal dependency of identity',[92] this is precisely how it is presented in *S&S*[93] (though this is unhelpfully dropped in *S&SR*). He is also right that, as it stands,

this formulation does not obviously claim anything about individuative procedures or our ability to *grasp* identity propositions. Yet Snowdon is wrong to take
this formulation as the guide to the general character of **D**.[94] So much should be
clear from the analysis above. The passage on page 56, which Snowdon takes to
capture **D**, is not, in fact, a guide to Wiggins's overall project, and is not, by
itself, intended to claim anything about individuation. The passage is intended
only as a partial rendering of a doctrine that encompasses both epistemological
and metaphysical claims. It is the point at which Wiggins incorporates the sortalism he finds in our individuative practices into his account of identity. It is a part
of his search for a sufficient condition for identity.

Snowdon's mistake appears even more starkly when the precise context of
the quoted passage is considered. Wiggins writes:

When D is clearly disassociated from R, that which remains is this:

D: $a = b$ if and only if … [etc.][95]

The principle is labelled **D** but at this point Wiggins is aiming to exhibit the elements of **D** that differentiate it from **R** (the 'relativity of identity thesis'). Both **D**
and **R** are sortal theories of identity – in the same way as Catholicism and Angli-
canism are both forms of Christianity. Certain elements need to be showcased
for the distinction between the two to be drawn out. But just as a claim about
deference to the Pope fails to capture the general character of Catholicism, so
too the emphasis on the metaphysics of identity fails to capture the general character of **D**.[96] Snowdon's attempt to capture the general character of **D** abstracts
important, but not isolable, elements of that thesis.

Encouraged by the methodological musings voiced in Chapter 1, we might
wonder whether Snowdon is being drawn along by a particular undertow in
Anglophone metaphysics. Snowdon's disassembly of Wiggins's interconnected theories is symptomatic of the 'piecemeal' attitude, clearly countervail-
ing the connective analysis that we find in Wiggins's and Strawson's texts.[97]
Wiggins's work *is* dense, and the grain of his argument *is* sometimes hard to
find but perhaps this complexity is a necessary corollary of his connective of
analysis? We might hazard that connective analysis is found to be particularly
obscure in an intellectual climate so strongly shaped by the Russellian, scientific, 'piecemeal' methodology.[98] This may well be one reason for Snowdon's
misreading.

A second reason why Snowdon neglects the interconnectedness of identity
and individuation is that he has a particular – and divergent – interpretation of
Wiggins's use of 'individuation'. Snowdon understands 'individuation' to be
something like the detection, by seeing, hearing, smelling (etc.) of an object in
our environment. He calls this 'perception-based object-directed thought'.[99] Cor-
respondingly, he interprets Wiggins's 'thesis of the sortal dependency of individuation' as claiming that to focus on an object in this way (that is, to pick it
out perceptually) depends on picking it out under a (medium-level[100]) sortal.[101]

And this he rejects. His objection stems from a series of counter-examples, which reveal his particular understanding of 'individuate':

> There seem to be lots of examples where people can think of an object in ignorance of its sortal type.[102]

> ... someone could point at a creature flying in the air and ask 'what is that?'[103]

> ... Science fiction films provide examples where an object falls to earth, and the scientists ask 'what is it?'[104]

> ... I hear a rustle and wonder 'What is that?', where the focus of the thought is the noise maker. This kind of perceptual contact does not enable me to determine what basic sort the item belongs to, but it does enable me to think about it.[105]

As Snowdon understands individuation, it can, as a cognitive act, take place *before* the sortal question ('what is it?') can be asked:

> I wake up in a darkened room in need of a drink, so want to locate my glass of water.... I can, as we say, make out objects close by, and wonder of one of them whether it is my glass. Now, here my thought is not directed onto the object by any sortal conviction. The targeting, rather, enables the sortal *question* to be raised about a particular item.[106]

It is on these grounds that Snowdon argues against the so-called 'psychological' thesis (or 'sortal theory of individuation'). He thinks we can pick out an object and, having done so, try *then* to subsume it under a sortal. We need not have picked it out as a specific kind of thing in order to think about it. These comments are not obviously wrong but their critical force lies in Snowdon's reading of 'individuation'. That reading contrasts with the understanding of 'individuation' that Wiggins articulates in *S&SR*.[107] (Snowdon is arguing against what Imogen Dickie calls 'appropriation sortalism' where Wiggins is interested in something closer to her 'rich referential sortalism'. Her commendable piece 'The Sortal Dependence of Demonstrative Reference',[108] sets this distinction out in greater depth, and she argues – largely compatibly with Wiggins – for the latter position.)

For Snowdon, individuation is simply a matter of latching on to, or referring to, an object in one's environment. Is the same true for Wiggins? To judge by the quotation he gives from the *Oxford English Dictionary*, individuation involves picking an object out in such a way that it can be 'distinguish[ed] in its environment from other things of like and unlike kinds'.[109] This seems at odds with what happens in the example of the glass in the darkened room. In Snowdon's scenario there is not enough information for the percipient to distinguish the particular from other similar sized objects.

Furthermore, for Wiggins individuation involves picking an object out in such a way that it can be *re-identified* at a later stage. This is part of the content of his reference to 'spatio-temporal boundaries' ('to determine or fix upon *x* in particular by drawing its spatio-temporal boundaries').[110] Contrast again, Snowdon's case of the mysterious rustler in the bushes. One may turn one's attention to the rustling in the hedgerow, and even, perhaps, claim to see some single *thing* that rustles there (rather than, for example, a nest of sparrow chicks). But can one then claim to be able to reliably *re-identify* the 'noise-maker'? If it ceases and begins again? It seems not. What if some fox had come along and eaten the chicks, and the noise was the sound of it settling down for a post-prandial nap?[111] Wiggins's notion of individuation is more demanding than Snowdon's (a fact further evidenced by his comment in *S&SR* – that '[s]ingling out is the sheet-anchor for *information about particulars*'.[112] This is something that Snowdon pointedly overlooks ('I prefer ... to abandon his claim about picking out being the sheet-anchor of information'[113]) precisely because it fails to accord with his more minimal reading).

Consider too the case of the jug and the material mass that constitutes it. One may turn one's attention to the particular region in space that they jointly inhabit, but before it is determined *which* of the two things is under inspection, answers to identity questions will remain elusive. (The lump may, for example, survive squashing whereas the jug does not.) The pervasiveness of material coincidence (to be argued for in the pages below) suggests that the significance of this issue is widespread – and it affects Snowdon's own examples. When the scientists ask of the alien object *what is it?* are they talking about the alien artefact or the mysterious alien mineral that makes it up? That needs to be settled if they are to pick 'it' out again in the future.[114]

If Snowdon wants to undermine the sortal theory of individuation he needs to provide a case in which we individuate some *x* – in Wiggins's sense of individuate – without that *x* being subsumed under a sortal. Consider then another example. Imagine I find some hard, shell-like thing in my garden; I do not know if it is a fossil, or a discarded carapace, or a chrysalis, or simply a shiny piece of stone.[115] I decide to keep it and – as is my wont – put it out of sight, in a box in a drawer in my desk. The next day I open the box and inspect the object again. What could cause me to question the claim that it is the same thing? Very little. Have I not then managed to re-identify it, without knowing what sortal it falls under?

Wiggins could accept this story.[116] Nor would he deny that when we *do* hazard claims about what some *x* is, we run what he calls an 'epistemological risk' – a risk that sometimes pays off.[117] In fact, he will note, we often pick things out incorrectly. As Snowdon indicates, we need not know what sortal a thing falls under for us to be able to pick it out. Wiggins is not claiming otherwise, however. While Snowdon's own reading of 'individuation' is less demanding than Wiggins's, his reading of the sortalism at play is much stricter; Wiggins does not think that the thinker, as Snowdon puts it, '[is] required to classify *x* *under a certain sortal* in order to single it out'.[118] His view, rather, is that while we might pick out some *x* without knowing what kind of thing it is, we do not, ever, pick it out as a 'thing' or a bare object.[119]

When Wiggins writes that individuation requires 'the thinker's singling $x$ out as $x$ and as a thing of kind f such that membership in f entails some correct answer to the question "what is $x$?"'[120] (the quotation upon which Snowdon bases the above reading), he need not hold that we know what 'kind f' is. His thought is that we pick $x$ out as a *thing of some sort*, and cannot pick it out as a plain particular. This is of a piece with his claims about spatio-temporal continuity. Entities move through time and space in specific ways; a cat persists in a different way from a fish. There is, Wiggins says, no *bare* continuity – but what else could the notion of an *object* supply? These thoughts come to the fore when we attend to the issue of re-identification, which Snowdon overlooks. To re-identify $x$, it cannot just be picked out as a thing *simpliciter*; we must already have offered suggestions (a carapace, a fossil or something else) as to what it is, what its principle of continuity amounts to, and against these suggestions we stake our claims about identity.

It is not clear whether or not Snowdon himself thinks that we can re-identify an object *qua* object. There is ambiguity in his statement that 'it is not necessary to single out *under any sortal*';[121] this could either mean that we need not know what sortal a thing falls under (in which case, there is no disagreement between him and Wiggins), or it could mean that we can individuate some thing without thinking that it need fall under some sortal (in which case, Wiggins *would* disagree). The problem with the examples he gives is that he seems to be suggesting that we pick these things out as 'things' or 'objects' without thinking they fall under sortals – and, as noted, this means we have no guidance in trying to re-identify them.

Wiggins's sortal theory of individuation withstands Snowdon's objections. The counter-cases either misconstrue what it takes, for Wiggins, for some $x$ to be subsumed under a sortal or they misconstrue his use of 'individuation' which is stricter than Snowdon's and involves the real possibility of the subjects *re-identifying* the thing in question.[122, 123] It is worth noting, in closing, that ignoring re-identification will obscure the central reciprocal development of the sortal theories of *identity* and *individuation*. For what part of Snowdon's analysis of individuation involves identity or assessment of identity claims? Individuation involves being able to say whether or not an earlier item is *the same as*, or *identical with*, a later item – and we answer such questions by deference to the sortals under which the items fall (though we might not, at first, have a sure grip on what those sortals are).

\* \* \*

The **D** theory is – as Wiggins has it – a seamless web,[124] and one in which the impolitic reader may become caught and confused. The aim in this chapter has been to emphasize the different strands that constitute it and their connections. It has been shown how, in a way consonant with his descriptivist methodology, Wiggins develops a sortal theory of *identity* and *individuation* that attempts to articulate the reciprocity between those two concepts.

# Notes

1  Wiggins 2012: 1.
2  Wiggins 2001: 6.
3  It should also be noted, before starting, that though it is a necessary feature of his connective analysis that neither *identity* or *individuation* (or *substance*) concepts are prior, he writes that, given that the investigation must start somewhere, he begins by turning to *identity* (Wiggins 2001: 18–20). We will do the same.
4  Wiggins 2001: 2.
5  Wiggins 2001: 1.
6  Wiggins 2001: 5.
7  Wiggins 2012: 3.
8  Wiggins 1996: 227.
9  See Wiggins 2012 for the clearest presentation of these proofs.
10  Wiggins 2012: 4.
11  Wiggins 2012: 4.
12  Wiggins 2012: 4.
13  For doubts about such see, for example, Geach 1962, Noonan 1976 and Gale 1984.
14  There is another feature of identity which Wiggins examines – that which we might call the independence of identity. He writes that

> if $x$ is the same thing as $y$, then their identity will hold regardless of how matters stand with other things.... The identity of $x$ and $y$ may of course be discovered in all sorts of ways that involve reference to other things, but constitutively the identity of $x$ and $y$ involves only $x$ and $y$'.
>
> (Wiggins 2012: 4)

This is formalized, in *S&S*, in his principle $D(x)$ – the subject of Gale's criticism in 1984 – and subsequently refined in Wiggins 2012 by his 'Adequacy Principle'.
15  Wiggins 2012: §5.
16  Wiggins 2005: 443. See also Wiggins 1967: 1, 2001: 56, 2012: 5.
17  Wiggins 2001: 56–57.
18  Wiggins 2012: 6.
19  Williams offers a helpful gloss on Wiggins's critical position here in Williams 2006 (1124).
20  There are other objections to Leibniz's second principle; they are detailed in Wiggins 2012.
21  Wiggins 2012: 7.
22  'It will be everywhere insisted, moreover, that the singling out at time $t$ of the substance $x$ *must look backwards and forwards* to times before and after $t$' (Wiggins 2001: 7).
23  Wiggins 2001: 6.
24  Wiggins 2001: 6.
25  Wiggins 2001: 21:

> If somebody claims of something named or unnamed that it moves, or runs or is white, he liable to be asked the question by which Aristotle sought to define the category of substance: *What is it?*

Indeed, this is a question that must always be asked (see Wiggins 1967: 1).
26  Wiggins 2012: 8.
27  Wiggins 2001: 21.
28  It is for this reason that he says, 'natural languages furnish so many locutions in the quasi-attributive form: "*a* is the same what as *b*?", "the same donkey, same man, same ..."' (Wiggins 2005: 444).

29  Wiggins's use (as described in 2001: 9, 77) is shaped by Strawson's, in Strawson 1959: 168–169.
30  For Wiggins's description of this distinction see Wiggins 2001: Chapter 1, §3, especially page 30 (also summarized in Snowdon 1990). Wiggins also occasionally talks of 'ultimate' sortals (as in Wiggins 1967: 32) – but this phrasing has been all but expunged from his recent work (though n.b. 2001: 129) and will not be discussed here.
31  Wiggins 2001: 30.
32  Wiggins 2001: 33.
33  Wiggins 2001: 56–57 (my emphasis).
34  Wiggins 2001: 57.
35  See, for example, Wiggins 2001: 56, 1967: 1.
36  Wiggins 2001: xii, 2012: 8.
37  Wiggins 2012: 9.
38  Wiggins 2012: 1.
39  Or an 'operation' (in the case of organs). Wiggins 2001: 72, 86.
40  Wiggins 2012: 8.
41  Wiggins draws the connection with Leibniz in Wiggins 1979: 313–315 and in Wiggins 2001: 84–85.
42  Wiggins 2001: 86.
43  Wiggins 2001: 84.
44  The ideas we grasp are

> ideas which are clear, as [Leibniz] would say, that is operationally effective for reliable application of a thing-kind word to a thing and yet indistinct, that is inexplicit, incomplete and open to improvement.
>
> (Wiggins 2012: 8, 2005: 444)

See also 2001: 83n.6.
45  Wiggins 2001: 59.
46  Wiggins 2005: 444.
47  Wiggins 2005: 444.
48  The status of **D**(i) is slightly obscure to me. It stands as a logical truth here (and in Wiggins 2001: 'it seems certain … that for each thing that satisfies a predicate such as 'moves', 'runs' or 'white', there must exist some known or unknown, named or nameable, kind to which the item belongs' (2001: 21)) – but if there is some proof, or argument other than the one from observation, for **D**(ii) it is not obvious what it is. And indeed, Wiggins deletes the passages relating to this in his later work (compare 1980: 62–68 and 2001: 64–69).
49  Wiggins 2001: 70.
50  Wiggins 2001: 69.
51  Can we really not individuate under more generic sortals? Does *animal* not encapsulate a general biological activity sufficient for saying whether *a* is identical with *b*? No. Think of the dramatic kinds of metamorphoses some organisms can survive and others cannot (compare sparrows, dogs, frogs and butterflies). One size does not fit all when it comes to animal individuation. But what about '*mammal*'? Could one re-identify some *x* by thinking of it as a mammal? Perhaps, but there are two reasons for doubt. First, there is still striking variation among the living activities of mammals (compare dogs, dolphins and duck-billed platypi). Second, there is surely something important (for the descriptivist, at least) in the fact that we do not, in practice, pick things out primarily as 'mammals'. Children, certainly, seem to grasp what it is to be a dog or a cat, before understanding the more generic sortal *mammal*.
52  Wiggins 2001: 72.
53  Wiggins 2001: 22, 27.
54  Wiggins 2001: 59.

55  Wiggins 2001: 72.
56  Wiggins 2001: 73.
57  Wiggins 2005: 445.
58  Wiggins 2001: 96.
59  Wiggins 2012: 22n.7.
60  Gale 1984.
61  Gale 1984: 240.
62  Wiggins 2012: 4.
63  Wiggins 2012.
64  Wiggins 2012: 4.
65  Wiggins 2001: 23.
66  See also Wiggins 2001: 140, and 1996 'Replies': 227.
67  Geach 1962, 1973, Noonan 1976, 1978.
68  Wiggins 1996 'Replies': 227.
69  E.g. Wiggins 1967: 32.
70  Wiggins 2001: 30.
71  It is to this distinction that theorists typically turn to explain what is meant by 'fundamental'. See Snowdon 1990: 4. See also Olson 1997: 28.
72  '[A]nd the jug and the coffee-pot cannot be construed as *phases* of an individual collection of matter' (Wiggins 2001: 34, 36–43). It is possible, indeed, that Wiggins would consider this construal of the situation as some form of 'process philosophy' – to which he is also opposed (see the end of Chapter 7, below).
73  Wiggins 1968.
74  Wiggins 2001: 36. The 'is' of *constitution* functions in the same way as the 'is' in 'The soufflé you are eating is flour, eggs and milk'.
75  Wiggins 1968: 90.
76  Wiggins 1968: 93.
77  These puzzles are presented clearly and succinctly in Olson 2009: 49f.
78  E.g. Rudder Baker 2000 and 2002.
79  Wiggins 2005: 444.
80  Snowdon 2009: 254.
81  Snowdon 2009: 255.
82  Snowdon 2009: 255.
83  Snowdon 2009: 271.
84  In understanding this, I have had considerable help from Wiggins 2012 (which was denied Snowdon in Snowdon 2009).
85  See, for example, Wiggins 1996: 244 (discussed in Chapter 3).
86  Wiggins 2001: xii.
87  Wiggins 2001: 19.
88  Wiggins 2001: 19–20.
89  Joll 2010: §iii.
90  Snowdon 2009: 262.
91  Snowdon 2009: 254–255.
92  Snowdon 2009: 254, fn. 1.
93  Wiggins 1980: 51.
94  He seems, indeed, to recognize as much; this appears on page 260 (and on page 257, where he notes that the third clause imports a considerable amount of conceptual baggage, thus presenting the fragment as part of a broader argumentative trajectory).
95  Wiggins 2001: 56.
96  Many thanks to Jennifer Hornsby for helping me precisify this point.
97  See, for example, Glock 2003: 165, where he also identifies Stuart Hampshire – another important influence on Wiggins (e.g. 2001: 195, fn. 3) – as a 'systematic' philosopher, rather than a piecemeal one.

98   This, indeed, is Strawson's own analysis of Wiggins's project. In his critical notice of *S&S*, Strawson writes (1981: 603):

> Much of the value of the book stems from its author's explicit recognition of 'how much can be achieved in philosophy by means of elucidations which use a concept without attempting to reduce it, and, in using the concept, exhibit [its] connexions' with other 'established and ... collateral' concepts (p. 4). It is such a procedure of connective, rather than reductive, analysis which fruitfully governs his treatment of the central concepts of identity, individuation, substance, sort, natural kind and essence.

99    Snowdon 2009: 264.

100   Snowdon 2009: 265.

101   Snowdon 2009: 262.

102   Snowdon 2009: 265.

103   Snowdon 2009: 265.

104   Snowdon 2009: 266.

105   Snowdon 2009: 266.

106   Snowdon 2009: 266.

107   There are other, more dramatically divergent understandings of 'individuation' (see Lowe 2003, and Wiggins 2001: 6, fn. 4).

108   Dickie 2011.

109   Wiggins 2001: 6.

110   Wiggins 2001: 6.

111   It is important that while, in some ways, Wiggins's position corresponds to Dickie's 'rich referential sortalism', his notion of individuation – which requires that the individuated object, if truly individuated, is *re-identifiable* once lost from view – is more demanding. She does not try and defend this sort of claim about sortalism – but it does not seem to me to be immediately inimical to her position.

112   Wiggins 2001: 6.

113   Snowdon 2009: 264.

114   Dickie has some interesting things to say on this topic (in Dickie 2011) which correspond to certain points that Snowdon makes in his essay. The case of the statue and the lump is what she calls a story of 'ambiguous pointing' (the fuller form of which she finds in Quine – in his paper 'Identity, Ostension, and Hypostasis' and in Wiggins's 1997, 'Reply to Fei Xu'). Quine argues that this kind of ambiguity in pointing can support a form of sortalism. Dickie disagrees – at least insofar as it is a form of 'appropriation sortalism'. Arguing from empirical evidence about visual attentiveness, she points out that we split the visual field into visual objects *prior* to conceptual thought. The boundaries that people like Quine think sortal concepts draw, are already in place:

> When you attend to a visual object there is a fact of the matter about the spatio-temporal spread of your attention. For example, if you are attending just to the arm of the chair, you will be quick to notice changes occurring on the chair's arm, and slow to notice changes on the chair's leg.
>
> (2011: 41)

> Similarly, it is if, but only if, you are attending to the piece of metal of which the chair is made, rather than to the chair itself, that you attentional link will remain intact through a change that the piece of metal but not the chair survives.
>
> (2011: 41)

At least with respect to instances of *complete* material coincidence I am less sure than Dickie. Does the attentional link remain 'intact'? What does that mean? Or, when the change occurs, does one not reassess what one is looking at – and, at that point, decide that one was looking at the lump of metal, supposing that it survives the crushing of the chair? Either way, I think that the point concerning the statue and the lump, is intended to support what Dickie calls 'rich referential sortalism'. Attentional links will only remain intact if you are constantly attending to the object. Wiggins is interested in our ability to pick things out in such a way that we can re-identify them (even after they have gone out of our visual field).

115 This story was proposed to me by John Dupré.
116 See, for example, his comments in the introduction to his latest collection (Wiggins 2016) about some small object, of some unrecognizable kind, which is kept on the mantelpiece and scarcely changes at all. On this basis he can be sure that, *whatever* it is, he can give that same thing back to its owner.
117 Wiggins 1995: 5.
118 Snowdon 2009: 262.
119 See Wiggins 1997 for clarity here.
120 Wiggins 2001: 7.
121 Snowdon 2009: 265.
122 Why does he do so? It is plausible that he attributes too much importance to the remark (in *S&SR*) that individuation is an act 'at a time'. He writes: 'Wiggins explains that he wishes to advance a characterization of 'what it amounts to, practically and cognitively, for a thinker to single a thing out at a time' (Snowdon 2009: 261–262). But while the cognitive act of singling out is undoubtedly *at a time*, Wiggins's interpretation of 'individuation' also explicitly involves the ability 'later to single out that same thing *as* the same thing' (Wiggins 2001: 1). Once again, for Snowdon, 'individuation' may be a cognitive act of isolating an object in experience, while for Wiggins 'individuation' provides for the possibility of the inquirer's *re-identification* of that object and his readiness to trace it through time (Wiggins 2001: 6).
123 Note, in connection with this, that Wiggins claims that to single out an *x* distinctly in one's environment necessarily *involves* being able to re-identify it and trace it through time. See Wiggins 1963: 189–90, 2001: 71, 72.
124 Wiggins 2001: 2.

# Bibliography

Dickie, I. (2011) 'The Sortal Dependence of Demonstrative Reference', *European Journal of Philosophy* 22(1): 34–60.

Gale, R. (1984) 'Wiggins's Thesis D(x)', *Philosophical Studies: An International Journal for Philosophy in the Analytic Tradition* 45(2): 239–245.

Geach, P. (1962) *Reference and Generality* (Ithaca: Cornell University Press).

Geach, P. (1973) 'Ontological Relativity and Relative Identity', in Milton K. Munitz (ed.) *Logic and Ontology* (New York: New York University Press).

Glock, H.-J. (2003) 'Strawson and Analytic Kantianism', in H.-J. Glock (ed.) *Strawson and Kant* (Oxford: Oxford University Press).

Grandy, R. (2008) 'Sortals', in E.N. Zalta (ed.) *The Stanford Encyclopedia of Philosophy* available online at http://plato.stanford.edu/archives/fall2008/entries/sortals/

Joll, N. (2010) 'Contemporary Metaphilosophy', *Internet Encyclopedia of Philosophy*, available online at www.iep.utm.edu/con-meta/.

Lowe, E.J. (2003) 'Review of *Sameness and Substance Renewed*', *Mind*, New Series, 112(October): 448.

Noonan, H. (1976) 'Wiggins on Identity', *Mind* 85.

Noonan, H. (1978) 'Sortal Concepts and Identity', *Mind* 87.

Olson, E. (1997) *The Human Animal: Personal Identity Without Psychology* (Oxford: Oxford University Press).

Quine, W.V.O. (1980) *From a Logical Point of View* (New York: Harper and Row).

Rudder Baker, L. (2000) *Persons and Bodies: A Constitution View* (Cambridge: Cambridge University Press).

Rudder Baker, L. (2002) 'On Making Things Up: Constitution and Its Critics', *Philosophical Topics* 30: 31–52.

Snowdon, P. (1990) 'Persons, Animals and Ourselves', in C. Gill (ed.) *The Person and the Human Mind* (Oxford: Clarendon Press).

Snowdon, P. (2009) 'On the Sortal Dependency of Individuation Thesis', in H. Dyke (ed.) *From Truth to Reality: New Essays in Logic and Metaphysics* (London: Routledge).

Strawson, P.F. (1959) *Individuals: An Essay in Descriptive Metaphysics* (London: Methuen).

Strawson, P.F. (1981) 'Review of *Sameness and Substance*', *Mind*, New Series, 90(360): 603–607.

Wiggins, D. (1963) 'The Individuation of Things and Places' (a Symposium with M.J. Woods), *Proceedings of the Aristotelian Society*, supplementary volumes, 37: 177–216.

Wiggins, D. (1967) *Identity and Spatio-Temporal Continuity* (Oxford: Blackwell).

Wiggins, D. (1979) 'Mereological Essentialism: Asymmetrical Dependence and the Nature of Continuants', in E. Sosa (ed.) *Essays on the Philosophy of Roderick Chisholm* (Amsterdam: Grazer Philosophische).

Wiggins, D. (1980) *Sameness and Substance* (Cambridge, MA: Harvard University Press).

Wiggins, D. (1982) 'Heraclitus' Conceptions of Flux, Fire and Material Persistence', in M. Schofield and M. Nussbaum (eds) *Language and Logos: Studies in Ancient Greek Philosophy presented to G.E.L. Owen* (Cambridge: Cambridge University Press).

Wiggins, D. (1995) 'Substance', in A.C. Grayling (ed.) *Philosophy 1: A Guide Through the Subject* (Oxford: Oxford University Press).

Wiggins, D. (1996) 'Replies', in S. Lovibond and S. Williams (eds) *Essays for David Wiggins: Identity, Truth and Value* (Oxford: Blackwell Publishing).

Wiggins, D. (1997) 'Sortal Concepts: A Reply to Fei Xu', *Mind and Language* 12: 413–421.

Wiggins, D. (2001) *Sameness and Substance Renewed* (Cambridge: Cambridge University Press).

Wiggins, D. (2005) 'Reply to Bakhurst', *Philosophy and Phenomenological Research* 17(2): 442–448.

Wiggins, D. (2012) 'Identity, Individuation and Substance', *European Journal of Philosophy* 20(1): 1–25.

Williams, S.G. (2006) 'David Wiggins', in N. Goulder, A.C. Grayling and A. Pyle (eds) *The Continuum Encyclopedia of British Philosophy* (London: Thoemmes Continuum).

# 3    Natural substances and artefacts

In Wiggins's view, judgments of identity or persistence demand that the entity or entities in question be individuated. What is *x*? What is *y*? Wiggins often says that the answer is given by the use of some substance term or (more generally) a sortal term. But not all sortal terms have the same standing or fulfil equally well Wiggins's various **D**-requirements. Some, such as 'student' and 'poet', make reference to the *phases* of a being. The usefulness of these terms for individuative purposes depends on the identification of the underlying kind that the thing or things more fundamentally belong to. Other sortal terms may denote kinds of entity that have been accorded a status arguably inferior to that of *substance*. What is at issue here?

For Wiggins, it appears that natural things, specifically *living things* – plants, animals and so on – are substances *par excellence*. He writes:

> [They] exemplify most perfectly and completely a category of substance that is extension-involving, imports the idea of characteristic activity, and is unproblematic for individuation.[1]

As he sees it, the semantics of *natural kind words* are such that they clearly encapsulate a specifiable *principle of individuation* for their members (as required by **D**(v)).[2] These terms which we use to pick out natural things – cats, dogs and so on – are semantically such that they refer to exemplars of that kind and the specific principle of activity its members exhibit.

There are, however, other objects in our environment that seem to lay claim to 'substancehood', and a correlative aim of Wiggins's work is to examine whether or not *artefactual* kind terms – 'table', 'car', etc. – can function as substance sortal terms as well. In the second half of this chapter, Wiggins's concerns with artefacts are discussed. He questions whether or not they realize genuine principles of activity. The aim in the final part is to assess his elucidation of the *substance* concept, and to examine whether he takes artefacts to be substances. Lynne Rudder Baker, Massimiliano Carrara, Peiter E. Vermaas and Michael Losonsky deny that he does. Their readings are scrutinized, found to be wanting, and an alternative line of interpretation developed.

## 1  The semantics of natural kind words

Before discussing the semantics of natural kind terms, however, it is necessary to distinguish, if only vaguely, between natural and artefactual kinds. The line has been drawn in different ways, but a prominent view – exemplified below in Rudder Baker's analysis – is that 'natural' things are things that are *independent* of human practices. 'Artefacts', by contrast, are taken to be products of human practices, fabricated by humans ('man-made') – and artefactual kinds are marked out by us and defined in relation to the practices from which they issue and the functions they perform. For what follows, this is labelled the *standard view* (stately more precisely below).[3]

How does Wiggins take 'natural kind' words, like 'cat' and 'tree', to *refer*? His discussion in *S&SR* begins with a critique of the 'nominalist essentialist' account of natural kind words. The nominal essentialist seeks, as Wiggins writes, 'to specify the sense of "sun" or "horse" or "tree" by a description of such things in terms of manifest properties and relations or in terms of appearance'.[4]

For example, the essentialist will group fruit together as *lemons* because of their yellow appearance, tartness of taste, thickness of skin (etc.). This is to play the role of taxonomist. And, as Wiggins points out, this way of selecting members of a kind is not *open to the world*.[5] Accounts that describe how natural kind words refer in terms of a *nominal essence* cannot explain how our conceptions of a kind can *evolve* (as we learn new things about them) while still being conceptions of that same kind. When scientific inquiry reveals to us e.g. that lemons have a distinctive genetic structure (if in fact they do) we will want to exclude from the kind *lemon*, fruit that are only superficially similar (like etrogs), and include bruised, unripe and discoloured fruit that have the appropriate molecular make-up. Thus:

> [A] more satisfying account [of natural kind semantics] will emphasize the contribution that the world itself makes to those conceptions.[6]

It is exactly this kind of receptivity that is supposed to be captured by Wiggins's *deictic-nomological* method,[7] a position built firmly on the foundations laid by Saul Kripke and Hilary Putnam.[8] As we find it in *S&SR*, this doctrine states that the explanation of the sense of a natural kind word revolves around *exemplars* (or 'stereotypes') of that kind (thus requiring 'context' or 'deixis') and the nomological (or 'law-bound') connections that hold between them. So, Wiggins writes:

> *x* is an f (horse, cypress tree, orange, caddis-fly … if and only if, given good *exemplars* of the kind that is in question, the best theoretical description that emerged from collective inquiries into the kind would group *x* alongside these exemplars.[9]

That is, there are law-like principles, known or unknown,[10] that hold between exemplars of a kind, and can thus collect together the extension of the kind around these representatives.[11] These gradually evolving theoretical descriptions (which Putnam calls the 'sameness relations') are the ones that, for example, biologists lay out as they investigate the natural world *a posteriori*.[12] The theoretical description encapsulates the 'principle of activity' to which we turn when individuating natural substances. It encapsulates the 'determinate pattern of growth and development towards, and/or persistence in, some particular form'[13] – and thus meets the central **D**(v) requirement for substance sortal terms.

There are nuances that, for the sake of succinctness, can be passed over here.[14] The core claim is that, on this model of reference, it is *natural kind words* which, *par excellence*, satisfy the requirements that flow from Wiggins's **D**-principles. He writes:

> [The *deictic-nomological* method] contains most of the answer to the problems that we have posed about the demands of **D**(iii), **D**(iv), **D**(v) and all the other **D** principles. If there have to exist true law-like principles in nature to underwrite the existence of the multiply instantiable thing that is the reference of a natural-kind predicate if law-like principles of this kind have to exist in order for that general thing's extension to be assembled around the focus of actual specimens or for a reality-invoking kind of sense to be conferred on the term standing for the concept f, then they must also determine directly or indirectly the characteristic development, the typical history, the limits of any possible development or history, and the characteristic mode of activity of anything that instantiates the kind.[15]

On the deictic-nomological model, natural kind words work well as substance sortal predicates because they refer us to the rich inner workings of the kind's members, and thus encapsulate the laws that constrain their persistence. (At this point, alarm bells will be ringing for philosophers of biology for whom 'natural kind' is a deeply controversial notion; I would ask that the alarm be muted for now – these issues will be addressed in Chapters 6 and 7.)

## 2  Artefact words and puzzles

On the *standard view*, natural items are contrasted with *artefacts*, objects like hammers and clocks, which are dependent on human practices and intentions. Such things are also subject to our individuative practices; we track them, assert ownership of them and so on. Yet, unlike Putnam, Wiggins rejects the application of the deictic-nomological method to such things on the grounds that the correlative terms lack nomological grounding.[16]

Consider – as Wiggins does – the case of a clock.[17] Clocks can be constituted of different materials and can work according to vastly different mechanisms (compare, for example, a sundial, a grand-father clock and a fob-watch). The nomological claims we can make about members of the kind *clock* are strikingly meagre in comparison to the copious and detailed biological and chemical descriptions that link members of, for example, the kind *human being*. The stereotypes lack internal or scientific resemblances, and cannot be grouped by reference to a common constitution. There are no hidden depths to plumb, so to speak, when asking whether *x* is a member of the kind *clock*.[18] Consequently, Wiggins contends, things like clocks can only be grouped under functional descriptions that are precisely *indifferent* to specific constitution and any particular mode of interaction with the environment.[19] In short, he sees members of the kind *clock*, and similar 'artefactual' kinds, as needing to be collected by reference to a conceptually shallow *nominal essence* – for example, a tin-opener is any instrument made for opening soldered tins, a pen is any ink-applying writing implement.[20] The semantics of these kinds of kind words make no reference to law-like dispositions or typical histories of their membership, but to a *function* – and it is to this (formalized into the 'principle of functioning in **D**(v)) that Wiggins directs us when individuating these kinds of things.

> [O]rdinary artefacts are individuated by reference to a parcel of matter so organized as to subserve a certain function.[21]

On one reading then, it seems that Wiggins sees artefact kinds to be grouped by reference to nomologically shallow functional essences, and that it is by reference to the matter that subserves this function that we individuate these kinds of thing.[22] Further, he seems to hold that the persistence conditions encapsulated by artefactual kind terms are in some sense more problematic than those of their natural kind counterparts. For example, the kind term *clock* alludes only to a particular function, not to a particular organization or continuity – and as a result does not rule against, for example, *disassembly*, *part-replacement* and *pauses of indeterminate length*. This shallowness is illustrated clearly by Wiggins:

> A clock may stop because it needs winding up. Such a pause does not prejudice its persistence. A clock can stop because it needs to be repaired; and again it persists, however long the lapse before the repair.... The nominal essence of *clock* must involve a stipulation of some sort concerning the capacity to tell the time. But surely the uninterrupted continuance for all *t* of the capacity at *t* to tell the time at *t* will not be stipulated. This is too strong. The only loss that could count to any appreciable degree against the persistence of the clock is a *radical and irretrievable* loss of the time-keeping function.[23]

This analysis leads Wiggins to the thought that the *semantic* difference correlates with a *metaphysical* distinction. The nomological shallowness provokes puzzles of identity which demonstrate a contravention of **D**(vi) (and, as noted, also creates a problem with **D**(x)).

One of the most notorious of these puzzles – one repeatedly discussed by Wiggins – is the so-called 'Theseus's ship' case. In this fission narrative, drawn from Plutarch's *Lives* by Hobbes and redeployed in *ISTC* and its sequels, we are asked to imagine a ship that has all of its parts gradually replaced with new planks, screws, etc. There is nothing in the nominal essence of *ship* that precludes this (nor would we want there to be). But imagine that someone collects all the old, discarded parts – blemished, but not unusable – and builds with them another vessel of exactly the same design as the first. Given the weakness of the conditions for artefact persistence – admitting both part replacement and disassembly and reassembly – the second vessel, made from the original parts, can *also* be seen to be the same ship as Theseus's. Here, then, is the puzzle: both resultant vessels can be construed as identical with the original, and yet they can hardly be identical with each other. The transitivity of identity is undermined (thus the contravention of his **D**(vi)), (and since the creation of the second ship appears to 'destroy' the original identity, **D**(x) is also contravened).[24]

In the situation described, the term *ship* fails to fulfil at least one condition that, for Wiggins, a term will meet if it is a substance sortal term.[25] The repercussions are potentially severe. We will begin to question whether *ship* can be a substance sortal concept at all. The same is true for *clock, pen, tin-opener* and so on. If it cannot preclude this sort of 'fissioning', the principle of functioning does not seem enough by itself to successfully pick out and trace an object through time. The conclusion we are led towards is that artefactual sortals are not sufficient for individuation. This, as Wiggins notes, correlates to the disquieting thought that the items these artefactual terms refer to are not 'genuine entities'.[26] Is this a view he seriously endorses? We will see.[27]

## 3 The *existence* of artefacts?

There are two main lines of interpretation of Wiggins's metaphysical analysis of artefacts. Massimiliano Carrara and Pieter E. Vermaas hold that Wiggins denies the *existence* of artefacts. Lynne Rudder Baker and Michael Losonsky claim that he considers artefacts to be, in some way, ontologically *inferior* to natural items, and not 'genuine substances'. Let us consider these complaints in this order.

Does Wiggins deny that artefacts, like chairs and tables, actually exist? This is the view Carrara and Vermaas attribute to him in their paper 'The fine-grained metaphysics of artifactual and biological functional kinds'.[28] Examining Wiggins's discussion of artefacts, they claim both that he denies artefacts are members of real kinds, and that this leads him towards an 'Aristotelian' conception, whereby 'metaphysically there are no such things as cars and tables because, in an Aristotelian vein, cars and tables do not have their own essences or principles of activity'.[29] This, they describe as an 'Aristotelian anti-realistic conception of artifacts'.[30] Setting it out in greater depth, they write:

> [W]e cannot find regularities in behaviours and form in functionally characterized artifacts, such as clocks, and they are not subject to common laws comparable to the natural kind case. Functional descriptors thus do not refer to an inner constitution of artifacts.
>
> The result is twofold. From an epistemological point of view the conclusion is that artifact kinds do not support induction at all. We cannot infer, for example, any truth about chairs from the observation of some instances

of chairs.... From an ontological point of view the result is that artifacts exist only in what Sellars (1963) calls the 'manifest image'. People project artifact careers, but by a serious ontological inventory of the world, *artifacts do not exist.*[31]

This passage is notable in two respects. Firstly, it attributes to Wiggins a dramatic, not to say melodramatic, view about chairs and tables: they do not exist. Secondly, it exhibits a peculiar overlap of metaphysical traditions: questions are being asked that are out of place in this methodological context; different strands of Western metaphysics are becoming snarled in the conceptual equivocation of 'substance' and 'existent'. To draw out this latter point first, it will be helpful to turn to some recent methodological studies by Jonathan Schaffer in which he turns to the critical, but neglected, distinction between Quinean and Aristotelian approaches to metaphysical practice.[32] Using these studies we can read Vermaas and Carrara's 'anti-realism' as a Quinean corruption of the Aristotelian position, and questionably appropriate as applied to Wiggins's work.

For Quine, as Schaffer sees matters,[33] the main task of metaphysics is to say *what exists*. Thus, the characteristic question with which Quine starts his inquiry: *what is there*?[34] In general, his answer is the understated 'everything', but he recognizes that there is room for disagreement over cases. We might wonder whether *properties* exist or *meanings* or *numbers*. It is the remit of metaphysics to determine which of these things exist.

The science-led method that Quine deploys in the pursuit of this task has already been mentioned. The only entities that we should admit to our ontology are those that are posited by our best theory.[35] For Quine (roughly speaking) that theory is physics. The theories by which we explain the physical interactions of matter are taken and translated into canonical logic. Those items that the bound variables must range over to be true are the items to be included on our ontological 'call-sheet'. (Thus the celebrated slogan: *to be is to be the value of a variable*.) One result of this kind of programme is that it produces what Schaffer calls a 'flat ontology'. There is no order to the list of things that exist. They are either included in the list or they are not. Schaffer writes:

> [T]he Quinean and the Aristotelian tasks involve structurally distinct conceptions of the target of metaphysical inquiry. For the Quinean, the target is *flat*. The task is to solve for E=the set (or class, or plurality) of entities. There is no structure to E. For any alleged entity, the flat conception offers two classificatory options: either the entity is in E or not.[36]

On the Quinean model, therefore, we are encouraged to ask: do artefacts exist? Counter-intuitive as it is the response, given the Quinean method, may be in the negative. For entities like chairs and tables (unlike quarks, perhaps) are inessential to physical explanations. (This is the picture – or at least part of the picture[37] – that we find in Peter van Inwagen's Quinean metaphysics ('there are,' he writes in *Material Beings*, 'no tables or books or rocks or hands or legs'[38]).)

In contrast to the flat ontology of Quinean metaphysics, the Aristotelian picture is of an *ordered structure*: a graded, hierarchy of being. As Schaffer reads him, Aristotle's attitude to the sorts of existence questions we find in Quine is trivializing.[39] Inquiring whether or not numbers,[40] time,[41] or the infinite,[42] exist his answer tends to be a dismissive 'yes'. He is concerned instead, with the further issue of *how* such things exist – or, more specifically, as Schaffer has it, his focus is on *ontological dependence*.

On the Aristotelian model, some things *depend on,* or are *ontologically posterior*, to others.[43] Consider, for instance, the relation between a mouth and a smile, or a quality like 'green' and a blade of grass. In these cases one thing is seen to depend upon another, not perhaps *causally* but (as will be discussed in Chapter 7) *essentially*. There are different ways in which items can ontologically depend upon others.[44] They can *inhere* in, be *posterior* to or be *grounded* on – and some of these will be examined in the pages below. But the important thought to focus on here is that, while the relation of ontological dependence is apparently absent from the Quinean programme,[45] that relation is the mainstay of the Aristotelian framework. One of the aims of the *Categories*, broadly speaking, is to mark out lines of dependence. A core issue for Aristotle – Schaffer contends – is determining what the *primary substances* are, those things that stand under (sub-stantia) the others, upon which the others depend.[46] *Substances* are the focus here, rather than existents.

These methodological musings may seem somewhat obscure, but the intermediary point is that Carrara and Vermaas's '*Aristotelian* anti-realistic conception', which denies the existence of artefacts, is a Quinean corruption of the Aristotelian claim (this is not surprising given the current dominance of the Quinean approach in the Anglophone world).[47] At least on Schaffer's reading, Aristotle would not have been concerned with whether or not to include artefacts in his ontology. Rather, the question that troubled him, and to which it appears he found a positive answer, was: are artefacts *ontologically posterior* to other entities?

This is the level at which Wiggins engages with the puzzles of artefact identity.[48] Nowhere does he ask whether artefacts *exist*.[49] Rather, his interest lies in whether the worries voiced above imperil the thought that artefacts are *substances*. Having outlined the Theseus's ship narrative, he remarks that it appears to drive us towards the 'fearful outcome ... anticipated in the high metaphysical tradition of substance that seeks ... to demote artefacts from the status of genuine entities'.[50] Clarifying this remark, he refers specifically to Aristotle, who 'maintained that natural things are the real beings *par excellence* to which everything else is secondary'.[51] He is not concerned that they might not *exist* (and who seriously is?); he is interested in whether they, like natural, living items, are *substances*.

## 4 The *substancehood* of artefacts

Ultimately, the Quineanism evident in Carrara and Vermaas' analysis might be rephrased in terms of issues about ontological dependence (indeed, Schaffer's contention is that such analyses presuppose these metaphysical relations).[52] Even so, would they be right to attribute any form of this Aristotelian view to Wiggins? He recognizes the issues and the possible demotion of things like clocks and computers, but does he deny their 'genuine' substancehood? And does this mean that he sees them to be ontologically *inferior* to natural substances? This is how both Losonsky and Rudder Baker read him. In 'The Nature of Artifacts',[53] Losonsky writes:

> David Wiggins defends the view ... that 'artificial machines' are not 'true substantial unities'.[54]

More explicitly, in 'The Shrinking Difference between Artifacts and Natural Objects',[55] Rudder Baker states:

> I am not claiming that Wiggins denies that there exist artifacts, only that he distinguishes between natural and artifactual kinds in ways that may be taken to imply the ontological inferiority of artifacts.[56]

In this section, the focus will be on Rudder Baker's analysis of Wiggins. On her reading, he holds that natural items are *genuine* substances and that artefacts are not, and are in some way ontologically *inferior* to them. Having constructed this reading she attempts to demolish it, stating that Wiggins's grounds for distinguishing artefacts from natural things either fail completely or fail to legislate any relevant ontological distinction between the two kinds. Her interpretation of Wiggins will be questioned and rejected.

Rudder Baker claims that Wiggins presents five possible ways of distinguishing between natural objects and artefacts, where the former can be conceived in some way ontologically 'superior' or more 'genuine' than the latter. They are:

1   Fs are genuine substances only if fs have an internal principle of activity.
2   Fs are genuine substances only if there are laws that apply to Fs as such, or there could be a science of fs.
3   Fs are genuine substances only if whether something is an f is not determined merely by an entity's satisfying a description.
4   Fs are genuine substances only if fs have an underlying intrinsic essence.
5   Fs are genuine substances only if the identity and persistence of fs is independent of any intentional activity.[57]

What is immediately striking is that, while (1)–(5) undoubtedly connect in some way to the condition that was found in *S&SR*, they represent a dismantling of Wiggins's original position.[58] Wiggins's view is that natural things are those that have principles of activity founded in law-like dispositions that form the basis for extension-involving sortal identification.[59] Thus, having a principle of activity (Rudder Baker's (1)) is intimately related to whether or not there can be a science – in the sense of an *a posteriori* investigation – of an item (2). Equally, the intrinsic essence of a substance is taken, by Wiggins, to encapsulate the principle of activity, so (1) and (4) do not seem to be separable either. It is not necessary to spell out all the connections between (1)–(5) in depth. The general point is that one might have initial concerns with another 'piecemeal' analysis of Wiggins's work.

We may also wonder when, exactly, the ontological judgement 'Fs are genuine substances only if ...' enters Rudder Baker's reading. Wiggins is focused on substancehood, but he talks of 'exemplifying the category of substance', not of being 'genuine substances' – and it is nowhere in her essay made explicit where this comes from.

These then are two preliminary concerns. They can be partnered with some further worries about Rudder Baker's rejection of (1)–(5). In her paper, she claims that Wiggins's distinction between artefacts and natural things is misguidedly grounded in one or more of (1)–(5). Her tactic is to go through each in turn and present counter-examples to show why these conditions fail to mark a genuine distinction, and why, consequently, they cannot support any ontological disparity between artefacts and natural items.[60] A brief survey of the counter-examples suggests, again, that her analysis misinterprets Wiggins's particular construal of notions like 'activity', and 'science', and 'essence'.

In refuting (1), she uses 'a heat-seeking missile'[61] as an instance of an artefact that possesses an internal 'principle of activity'. Yet it is by no means clear that this captures what Wiggins means by a 'principle of activity'[62] – that is, a nomologically grounded *mode of being* about which we may learn unknown and potentially surprising facts. Heat-seeking missiles lack nomological depth. There are related difficulties with Rudder Baker's counter-example to (2), where she offers 'computer science'[63] as a case where artefacts are the subject of scientific inquiry. Computer science, presumably, is not the kind of *a posteriori* enterprise that Wiggins is thinking of, which attempts to fill out the theoretical descriptions holding between exemplars (again, it seems unlikely that we will *discover* new facts about Amstrads, in the way that we might with a new species of frog).

Further, in rejecting (3), Rudder Baker offers 'gold' as an example of a natural thing[64] – but one may well question whether this is a *sortal* term, in the sense laid out above, since (like 'water' and 'air') it fails to take numerical modifiers. (Also in response to (3), she describes a situation where archaeologists believe two artefacts to be of the same kind, without knowing what they are (i.e. whether or not they were used in battle or in religious rituals), to substantiate the thought that artefacts can be determined indexically rather than by satisfying a description. Significantly, however, picking out these archaeological finds indexically will not allow the archaeologist to pick them out as artefacts rather than, potentially, *parts* of artefacts.)

Her reading of Wiggins's analysis of 'natural kind' and 'intrinsic essence' in (4) is also problematic. She denies that natural things necessarily have an underlying intrinsic essence, and cites wings (of birds and insects) as counter-examples to the fourth claim. This however, misconstrues Wiggins's view of natural kinds and natural substances – in anticipation of the arguments in

Chapters 8 and 9, one line of interpretation finds him rejecting the idea that organs or body parts register as substances.

With these concerns raised, an intermediate point can be made. As well as being ill-fitting, these listed counter-examples indicate that Rudder Baker has a *prior* understanding of the distinction between natural things and artefacts against which she is measuring Wiggins's putative conditions. She claims that the distinctions described in (1)–(4) fail because there are natural things and artefacts that are not accommodated by this suppressed view. It is in (5) that this prior understanding of the distinction is brought to the fore. She states that (5), unlike (1)–(4), *does* distinguish between artefactual and natural kinds.[65]

An artifact's being the kind of thing that it is depends on human intentions.[66]

And, elsewhere:

Artifacts are objects intentionally made to serve a given purpose; natural objects come into being without human intervention.[67]

For Rudder Baker 'artefacts' are intentionally made to serve a particular purpose.[68] Hers is a statement of, what was called above, the '*standard view*' of the distinction between natural things and artefacts – and she goes on to claim that Wiggins must turn to (5) to undergird his artefactual/natural distinction: the independence from human intentions determines the difference between artefacts and natural things. For Rudder Baker, the ontological disparity she finds in Wiggins can only be based on this fifth condition: artefacts are ontologically inferior because we made them; they depend, in some way, on the human mind.

Yet this critical method, and indeed the generalizing found in Vermaas and Carrara, fails to register the important point that Wiggins's actual condition is intended to be *stipulative*.[69] Wiggins demonstrably avoids endorsing the *standard view* – a point he makes explicit in the texts to which Baker, Vermaas and Carrara all refer. He states, pointedly, that his distinction does *not* map onto the distinction between fabricated objects which are the products of human minds, and natural objects which are not:

[A] particular continuant $x$ belongs to a natural kind, or is a natural thing, if and only if $x$ has a principle of activity founded in lawlike dispositions and propensities that form the basis for extension-involving sortal identification(s) which will answer truly the question 'what is $x$'? ... it is not the question of whether a thing was fabricated but rather the difference between satisfying and not satisfying this condition that makes the fundamental distinction. Loosely and because there is no other handy term, I shall continue to call objects that fail this crucial condition 'artefacts'. But this is without prejudice to the question ... of the possibility (which I have no wish to prejudge) of the artificial synthesis of natural things.[70]

He specifically recognizes the possibility of artificially (intentionally) synthesized natural objects.[71] He also accommodates artefactual readings of non-man-made objects such as wasp's nests or India rubber balls.[72] Similarly, he holds that some things, that register as artefacts on the *standard view* – like works of art[73] – have a nomological depth beyond that of, for example, chairs and are less 'artefactual' as a result. Rudder Baker's is a questionable interpretation of the distinction Wiggins draws. It may not matter very much that Wiggins's account fails to correspond to the *standard view*.

## 5 Substance

The following interpretation is guided by comments that Wiggins makes in his 1995 paper 'Substance'.[74] Ostensibly this is an introductory essay, but one of impressive complexity. It starts by explaining Aristotle's use of the term 'substance' (or its correlates *ousia, on hupokeimenon*, etc.) and situates it firmly within the tradition that Schaffer, as above, contrasts with the Quinean programme.[75] The essay goes on to respond to the empiricist rejection of the notion and in doing so it encompasses the issue of the substancehood of artefacts.

Wiggins states that among our core concepts, and alongside our notion of *sameness*, is the concept of *substance*. When we look to our everyday interactions we see that they are premised on the primitive idea of a 'a persisting and somehow basic object of reference that is there to be discovered in perception and thought.'[76] Here, I quote liberally:

> Salient among the things that we have to recognize, if we are to make sense of the world, are the substances.[77]
>
> There is a central thought in our conceptual scheme which we do not know how to do without, that we can gradually amass and correct a larger and larger amount of information about one and the same thing, the same subject, and can come to understand better and better in this way how these properties intelligibly cohere or why they arise together.[78]
>
> Their claim is a claim on our practical and theoretical reason. Everything conspires to force them upon us if we have the slightest concerns to find our way about the world or understand anything at all about how it works.[79]

Wiggins's descriptive approach elucidates the concept of *substance* by turning first to the primitive notion he finds underlying our practices. From the outset, therefore, the notion of substancehood is supposed to be attuned to the world-view of the human inquirer.[80] As is typical of the descriptive approach, there is a blurring here between ontological and epistemological issues:

> For us, the importance of the category of substance ... is not so much onto-logical as relative to our epistemological circumstances and the conditions under which we have to undertake inquiry. These circumstances and conditions determine where we have to begin in order to find our way about, in

order to designate spatial and temporal landmarks, and in order to find workable, dependable, low-grade generalizations about how identifiable classes of things come into being, persist and behave.[81]

We turn to how the *substance* concept functions in our thoughts to elucidate this metaphysical notion. As I have already emphasized, this elucidation is one strand in Wiggins's wider connective analysis, and proceeds in concert with elucidations of other central concepts, relating to *identity* and *individuation*. Some of these connections are set out above, and some of the elements of his elucidation of substance appear in the quotations below.

One, aforementioned, component of the concept is that the continuity of a substance cannot be understood as *bare* continuity. The entities around us are more than persisting 'objects'. It is part and parcel with the business of singling things out and tracking their course in space and time to pay attention to what they do and how they behave – their *principle of activity*.[82] A cat persists in a certain way, a mouse in another and a dolphin in another way too. Wiggins sees our pre-theoretical *substance* concept as encapsulating some notion of entities that 'have a source of change or principle of activity within them'.[83]

Another component of the pre-theoretical *substance* concept is, Wiggins thinks, highlighted by an analysis of grammar. Following Strawson and Aristotle, he focuses on the grammatical distinction between *subject* and *predicate* (the 'subject' being the element of the sentence (like 'Socrates') that refers to a particular thing, where the predicate refers to a general characteristic borne by that thing (e.g. 'is wise')).[84] For us to engage with the world in the way that we do we have to pick out items which submit to predication without being predicated – or so Aristotle leads Wiggins to declare.

> A substance ... is something that is neither in anything else nor predicable of anything else.[85]

Referring back to the discussion above, and viewing things through the descriptivist's lens, one can see how this grammatical distinction can be taken to exemplify some form of *ontological dependence*. It is a feature of our language, he says, that there are some items (substances, like humans) on which other items (properties, like smiles) depend. And this seems to be the thought expressed in the following passage from 'Substance' (an exposition of the *Categories*, Chapters 1–5):

> Among the different subjects you can talk about, some are and some are not in others in the way in which colours and their determinate shades are in things. Some are and some are not in things in the way in which knowledge in general or some specific knowledge ... is in things.... To the extent that anything is not *in* other things in this way, it enjoys a certain autonomy.[86]

Herein, Wiggins claims that we see some things – colours and so forth – to be *in* other items, to depend in some manner upon them. *Substances* are those things upon which other things depend (and here, at least, 'ontological dependence' seems to be understood as a basic relation in our conceptual framework). The worries with conceptual invariance are pressing here, especially given Burtt's and Mei's critiques of Strawson's focus on this distinction – these will be addressed in Chapters 4 and 5.

A third central element which Wiggins isolates in the Aristotelian *substance* concept is that of 'internal cohesiveness'[87] or 'real unity'.[88] In our day-to-day dealings, we treat these basic objects as being *more* than collections of inter-changeable parts (in contrast, for example, to Theseus's ship). This thought (examined in greater depth in Chapter 8) can also be construed in terms of a type of ontological dependence;[89] we see certain things as being *prior* to their parts (we understand hearts, for example, by reference to the role they play as parts of the whole organism). In this connection, Wiggins quotes Spinoza:

> By a substance I mean that which is in itself and is conceived through itself: that whose concept makes no essential reference to anything else.[90]

Substances are, in some sense, complete and discrete. *Qua* substances, we deal with them as individual things, not as conglomerates, nor as parts of larger entities.

Here, then, are three central features of Wiggins's analysis of the *substance* concept: *activity*, *grammatical primacy* and *unity*. It is a cursory sketch but I hope that it provides a clearer sense of his project. His aim is to draw out the interrelated elements of the concept that underwrite our everyday practices, to understand the category that orders our thoughts and allows us to find our way about, pick out things and understand their persistence and behavior.

Significantly, Wiggins also thinks that as we bring this concept to bear on the world we find that some items realize these features to lesser and greater degrees. Some objects more obviously 'exemplify the category of substance'.[91] For Wiggins, it is not a 'yes/no' matter of either being or not being a substance (this much is recognized by Baker, who talks of 'inferiority', and it is missed in Vermaas and Carrara's binary picture).

> It will relax the intellectual cramp that threatens if, instead of trying to decide the question whether artefacts as a class are or are not substances, we resolve to reinterpret the question and see it as a question about the *distance* at which this or that particular artefact (or this or that group of artefacts) lies from the central case in respect of durability, internal cohesiveness, having a relatively self-contained principle of activity, and exemplifying some simple law of change.[92]

It is important to note that in this passage, from 'Substance', Wiggins is writing about artefacts as they are defined on the *standard view*. Contrary to Baker's assessment, Wiggins is in a position to maintain that artefacts (on the *standard view*) can exemplify the category of substance.[93] Nor should this be surprising,

Wiggins suggests, since fabricated objects are made in such a way that they register as basic persisting objects in our pre-theoretical framework.

> [T]he solidity, durability, and internal cohesiveness of a vast preponderance of our artefacts, some of them outlasting their makers (who certainly were substances) by millennia, would be a standing reproach to any would-be puristic ruling to the effect that artefacts stand at too great a distance from the natural continuants that furnished us with our original paradigm of substance. Indeed such a ruling would represent in at least one way an affront to the spirit of the original conception. For not only do artefacts submit to predication without being predicated, not only can they furnish us with a 'this' and furnish (insofar as we know what this means) something 'separable'. Their usefulness and effectiveness in the performance of the functions signals and celebrates the very same evolving understanding of the way ordinary perceptible things behave that made the notions of substance, of nature, and of substances with their natures so interesting and important to us in the first place.[94,95]

*Are artefacts substances?* The question is blunt and at once ill-suited to the subtle elucidatory analysis Wiggins offers of *substance* and neglectful of the distinction he draws between artefacts and natural items. *Are artefacts genuine substances?* This is better, but the binary distinction between genuine and non-genuine still fails to correspond to the spectrum of substancehood that Wiggins articulates. What then *is* his view? By definition, 'artefacts' do not have principles of activity but principles of functioning (or operation).[96] Depending on the complexity and nomological depth of the principle of functioning, artefacts may more or less clearly exemplify the category of substance – but definitionally they are less substantial than natural things, which possess a principle of activity.[97]

Perhaps, in some respects, Rudder Baker is right; Wiggins *does* see artefacts to be less substantial than natural things. Yet one must read this metaphysical pronouncement carefully. It is easy to take it to mean that items which fall further from the paradigm substances, and exemplify that category less well, are somehow less 'real' than natural things; this is what is suggested by Rudder Baker's talk of 'ontological inferiority'.[98] It is, however, an interpretation that sits awkwardly alongside what I have called the 'open-minded' aspect of Wiggins's project.

The quotations above indicate that Wiggins is prepared to see certain things – colours, for example – to be 'ontologically inferior' to the objects in which they inhere, insofar as 'inferiority' is read as 'posteriority'. Thus there is some kind of *hierarchy* in his metaphysics. And if he holds, as Rudder Baker believes he does, that artefacts are, as a class, *mind-dependent* on natural objects – dependent, that is, on us – there might be some basis for this thought. But he does not hold this. For Wiggins has no taste for desert landscapes. His metaphysics is verdant and fertile; therein we find *substances* – whose metaphysical character has been sketched above – and *properties* that inhere in those substances. And there are other entities besides, which are not 'posterior'/'inferior' to substances, yet have distinct metaphysical characters. Among them are 'concrete universals' (discussed in Chapter 9) and 'processual' beings (discussed in Chapter 7). Moreover, his metaphysics is not in thrall to the exclusionary 'ambitious' descriptivism that some find in Strawson.[99] He makes no attempt to discredit the structures or entities described in other metaphysical frameworks. He is hospitable to the thought that there are different metaphysical frameworks or models which partition reality in various ways – models that posit four-dimensional objects, or processes, or mereological simples, or aggregates rather than substances. Such models may be alien to one another,[100] but why should they not be *cotenable*?[101] Concrete universals, processes, four-dimensional objects and mereological simples may not be substances. But they need not be 'ontologically inferior' (in the sense of 'posteriority') or less 'real'.

Maybe artefacts do not exemplify the category of substance. Yet this should not prejudice us against their reality. At this point I make one more suggestion (to be developed in Chapter 9). The richness of Wiggins's metaphysical language allows him to go beyond this relatively uninformative pronouncement about substancehood. Wiggins has to ask: if artefacts are not paradigm substances, *what are they*? He has the resources to examine their metaphysical character in greater depth. He might point to their principle of functioning or how (like substances) they submit to predication. More importantly he can say that unlike substances, their parts are conceived as being ontologically *independent* of the whole (this latter feature explains why it is that we care as little as we do about artefactual part replacement).[102] This description of their metaphysical character will be taken up again in later chapters. For now the point is made. To be a substance is not a yes/no matter – nor does substancehood correspond to ontological superiority.

\* \* \*

It was shown, in Chapter 2, how **D** encapsulates interconnected thoughts about *identity* and *individuation*. The *substance* concept can now be seen to be another facet of this theory. The above exposition sets out Wiggins's account of this basic notion, its centrality in our individuative practices and how different items – aretefactual or natural – may exemplify it to lesser and greater degrees. This chapter has also identified and rejected variant readings of Wiggins's work, while simultaneously giving reasons for these variant readings.

Some points have also been raised for further investigation. Wiggins is seen to rely heavily on the viability of *conceptual invariance* (with respect to grammar), and this is discussed more fully in the next two chapters, with respect to the *person* concept. Questions were raised as well, about the distinction between artefacts and natural objects, and it was suggested that the metaphysical character of artefacts may be fleshed out in terms of the ontological dependence of their parts – this will bear on the discussions of *biological mechanism* and *transplantation* in Chapters 8 and 9. More work has yet to be done – so onwards.

## Notes

1  Wiggins 2001: 90.
2  Wiggins 2001: 72.
3  For a further analysis of the 'standard view' see Hilpinen 1993: 156–157.
4  Wiggins 2001: 78.
5  Wiggins 2001: 78, 160, 173.
6  Wiggins 2001: 78.
7  See especially Wiggins 2001: 79–80ff.
8  Kripke 1980 (and – though less directly influential for Wiggins – Putnam 1973).
9  Wiggins 2001: 79.
10  Wiggins 2001: 72, and 1980: 80f.
11  Wiggins 2001: 80.
12  Wiggins 2001: 86.
13  Wiggins 2001: 86.
14  Some of these subtleties will be discussed in Chapter 6. For example, we might raise questions about how the sameness relation is supposed to be identified in the first place, since – surely – weighting different areas of similarity differently will create different measures of 'sameness'. This thought, essayed in Okasha 2002, is also discussed below.
15  Wiggins 2001: 84.
16  Wiggins 2001: Chapter 2, §2–3. In this, Wiggins's position aligns with Schwartz (see Schwartz 1978). For a good introduction to these issues, and an overview of the debate before Wiggins's arrival, see Kornblith 1980.
17  It is this example with which Losonsky takes umbrage (in Losonsky 1990). He believes that artefacts *do, contra* Wiggins, develop in a nomologically grounded way.
18  It is for this reason, as Wiggins points out, that we are never surprised by facts about artefacts, but are constantly astonished by the intricate workings of members of natural kinds (of the order of surprise that might strike one on learning that tadpoles are frogs) (Wiggins 2001: 88).
19  Wiggins 2001: 87.
20  Wiggins 2001: 87.
21  Wiggins 2001: 91.
22  'Functioning', as Wiggins has it, is no more 'than remotely analogous to the activity of natural things' (Wiggins 2001: 90).
23  Wiggins 2001: 91–92. Wiggins continues: 'even under this circumstance, the clock itself may be held to have survived ...' – and we may wonder whether there are biological analogues of this. The living activity of seeds trapped – and preserved – in glacial ice might, for instance, be said to undergo a pause of indeterminate length (this is Jack Wilson's focus in Chapter 5 of his book, *Biological Individuality* (1999: 101f)). But in these cases, at least, this kind of stasis appears to be a *facet* of the living activity – it is part of the mode of being of such things that they can survive freezing. Contrast with the artefacts that survive *in spite of* such interruptions.
24  On this, see Wiggins 2001: 93, 99, where he configures this problem as a tension between the ordinary commonsensically strict notion of identity and the commonsensically loose requirements of artefact identity. Note, also, that these kinds of artefact puzzles are not applicable to all supposed artefacts. Artworks are especially interesting here, and Wiggins devotes some time to exploring the degrees of replacement and repair that, for example, paintings can undergo (Wiggins 2001: 136–139).
25  This failure can also be formulated in terms of **D**(iii) – see Wiggins 2001: 70 and 92.
26  Wiggins 2001: 99–100.
27  The focus above is on the general worries that puzzles like this provoke, rather than on Wiggins's specific response to this particular case. However, for the sake of completeness, a brief overview can be entered. Wiggins's thoughts about Theseus's

ship have changed. In *S&S* and *S&SR* he suggests we deploy supplementary principles to bolster artefact identity conditions. He adapts an identity condition for quantities, formulated by Helen Morris Cartwright. The – almost comically gestural (Wiggins 2001: 100) – condition concerns the addition and subtraction of matter: matter can be exchanged and replaced so long as (i) the artefact retains its capacity to perform the function for which it was designed, and (ii) it retains more than half its matter (ruling out fission). More recently, in his 2012 paper, Wiggins proposes a slightly subtler response: we can only answer this question when we have entered into serious dialogue with those who make and use the artefact:

> it would be wise for philosophy not to hold itself aloof from the uses of those who[se] ship it is. Typically, they will decide such a matter not once and for all but incrementally and in a way that the theorist needs to understand before he ventures to criticize.
>
> (Wiggins 2012: 15)

The suggestion, then, is that answering such puzzles is a bigger project than philosophers typically take it to be. It will be a matter of more thoroughly elucidating our thoughts about artefacts and ownership – and while no further verdict on the case will be offered here, the closer focus on the descriptive analysis of our *artefact* concept will be encouraged in Chapter 9.

28  Carrara and Vermaas 2009.
29  Carrara and Vermaas 2009: 126.
30  Carrara and Vermaas 2009: 126.
31  Carrara and Vermaas 2009: 130 (my emphasis).
32  E.g. Schaffer 2003, 2009, 2010.
33  Schaffer 2009: 347–348.
34  Quine 1963: 1.
35  Quine 1963: 12–13.
36  Schaffer 2009: 354.
37  Peculiarly, van Inwagen (1990) claims that organisms exist, and do so because they have a single biological life, which organizes them in such a way as to create unity. Here again we see a bizarre curdling of metaphysic. He claims his position is Aristotelian *and* Quinean; but being Quinean, he is reductionist about biology and, being reductionist about biology, and reducing biological life to a matter of physico-chemical reactions, he cannot also be Aristotelian in the sense (as he claims) of attributing privileged substancehood to organisms. For this reason, I have always been somewhat suspicious of the account offered in *Material Beings* – and it strikes me that this might be an instance of a philosopher's external (theistic) beliefs, affecting metaphysical arguments. The life principle in van Inwagen's work, seems at points to work very similarly to the soul.
38  Van Inwagen 1990: 18.
39  Schaffer 2009: 348.
40  Aristotle *Metaphysics* 1077b32–3.
41  See Owen 1986: 275.
42  Aristotle *Physics* 206b13–16.
43  For an overview of Aristotle's notions of *priority* and *posteriority* and how they relate to dependence, see Gill 1989: 3ff. This will be discussed in greater depth in Chapter 8.
44  See, for example, Correia 2008: 1013.
45  Schaffer argues that the Quineans have attempted to eschew talk of dependence, yet remain implicitly committed to it – see Schaffer 2009: §1.
46  This is not to say – as Wiggins makes clear in his paper 'Substance' (1995) – that Aristotle thought there could be 'bare substrata' of the kind that Hume so viciously mocked.

47  Schaffer 2009: 347.
48  It is interesting that the Aristotelian analysis of dependence is enjoying something of a revival at the moment. Philosophers like Schaffer and Fine are seen to be effecting a 'significant reorientation' (Koslicki 2012: 186) in the Analytic sphere, suggesting that questions in metaphysics are more profitably understood as questions about dependence. This is a shift we will use to our advantage in the discussions in Chapter 8.
49  Hark back to Snowdon's paper 'On the Sortal Dependency of Individuation Thesis': therein he accuses Wiggins of an 'identity fixation' (Snowdon 2009: 258) and suggests that sortal dependency of identity might better be rendered as the 'sortal dependency of *existence*' thesis. The above considerations suggest otherwise. Snowdon states that 'the crucial thesis is that *existence* depends on, or requires, *falling under a sortal*?' But, as mentioned, Wiggins's concerns are not with existence questions.
50  Wiggins 2001: 99–100.
51  Wiggins 2001: 100, fn. 25.
52  Schaffer 2009: §1–2.
53  Losonsky 1990.
54  Losonsky 1990: 81.
55  Rudder Baker 2008 (this paper is a reworking of her 2004).
56  Rudder Baker 2008: fn. 5.
57  Rudder Baker 2008: 3–4.
58  This is made explicit by Rudder Baker

> All the conditions either follow from, or are part of, the basic distinction that Wiggins draws between natural objects and artifacts. There is a complex condition that natural objects allegedly satisfy and artifacts do not: '...a particular constituent x belongs to a natural kind, or is a natural thing, if and only if x has a principle of activity founded in lawlike dispositions and propensities that form the basis for extension involving sortal identification(s) which will answer truly the question 'what is x?'.
>
> (2008: fn. 5)

59  Wiggins 2001: 89.
60  Rudder Baker 2008: 4–6.
61  Rudder Baker 2008: 4.
62  Wiggins 2001:

> [A] delicate self-regulating balance of serially linked enzymatic degradative and synthesizing chemical reactions [that enables an object] to renew [itself] on the molecular level at the expense of those surroundings, such renewal taking place under a law-determined variety of conditions in a determinate pattern of growth and development towards, and/or persistence in, some particular form.
>
> (86)

63  Rudder Baker 2008: 5.
64  Rudder Baker 2008: 5.
65  Rudder Baker 2008: 6.
66  Rudder Baker 2008: 6.
67  Rudder Baker 2008: 1.
68  Rudder Baker 2008: 1.
69  This is clear from Wiggins 1980 (89) and Wiggins 2001 (89) though, as discussed below, he examines the standard view in Wiggins 1995.
70  Found in Wiggins 2001: 89.
71  Wiggins 2001: 90 (also, for example, Wiggins 1996 'Reply to Snowdon', and Wiggins 1980, Chapter 6).

72  Wiggins 2001: 90.
73  Wiggins 2001: 136–139.
74  Wiggins 1995.
75  Wiggins 1995: 214ff.
76  Wiggins 1995: 214.
77  Wiggins 1995: 216.
78  Wiggins 1995: 216.
79  Wiggins 1995: 245.
80  Wiggins 1995: 217.
81  Wiggins 1995: 244.
82  Thus, '*[s]ubstance*, so understood, and *activity* … are notions made for one another' (Wiggins 1979: 315). See also Wiggins 1995: 218.
83  Wiggins 1995: 219.
84  See Strawson 1974 (see also Snowdon 1998/2009 for a commentary). The grammatical feature – which Burtt and Mei take to be culturally local – is understood to exemplify a central element of our conceptual scheme, and thus (by Strawson's descriptivism) to stand as an indication of metaphysical structure.
85  Wiggins 1995: 216.
86  Wiggins 1995: 216.
87  Wiggins 1995: 242.
88  Wiggins 1980: 98.
89  The connection between 'ontological dependence' and 'priority' is articulated well by Gill (in Gill 1989).
90  *Ethics*, first part, definition III, quoted in Wiggins 1995: 223.
91  Wiggins 2001: 90.
92  Wiggins 1995: 243.
93  Wiggins, for instance, will see intentionally produced, synthetic organisms and certain artworks, to be substances (Wiggins 2001: 89–90).
94  Wiggins 1995: 242.
95  Wiggins encourages us to think of the 'sense of dislocation that would result from our withholding the status of substance' from them (Wiggins 1995: 242).
96  Wiggins 2001: 89.
97  'Substances are things that have a source of change or principle of activity within them' (Wiggins 1995: 219).
98  I say 'suggested' because nowhere does she specify exactly what she means by 'inferiority'.
99  Haack 1978: 365f.
100  Wiggins 2001: 31.
101  Wiggins 2001: 155–156.
102  Wiggins recognizes this and further adds that

> [t]he truth is … that, for some practical purposes, we simply do not mind very much about the difference between artefact survival and artefact replacement.
>
> (Wiggins 2001: 101)

## Bibliography

Aristotle. (1936) *Physics,* W.D. Ross (ed. and trans.) (Oxford: Clarendon Press).
Aristotle. (1994) *Metaphysics* (Books Z and H), D. Bostock (ed.) (Oxford: Clarendon Press).
Carrara, M. and Vermaas, P. (2009) 'The Fine-grained Metaphysics of Artefactual and Biological Functional Kinds', *Synthese* 169: 125–143.
Cartwright, H.M. (1965) 'Heraclitus and the Bath Water', *Philosophical Review* 74: 25–42.
Cartwright, H.M. (1970) 'Quantities', *Philosophical Review* 79: 25–42.

Correia, F. (2008) 'Ontological Dependence', *Philosophy Compass* 3(5): 1013–1032.

Gill, M.L. (1989) *Aristotle on Substance: The Paradox of Unity* (Princeton University Press).

Haack, S. (1978) 'Descriptive and Revisionary Metaphysics', *Philosophical Studies* 35: 361–371.

Hilpinen, R. (1993) 'Authors and Artifacts', *Proceedings of the Aristotelian Society* 93: 155–178.

Kornblith, H. (1980) 'Referring to Artifacts', *Philosophical Review* 89: 109–114.

Koslicki, K. (2012) 'Varieties of Ontological Dependence', in F. Correia and B. Schnieder (eds) *Metaphysical Grounding: Understanding the Structure of Reality* (Cambridge: Cambridge University Press): 186–213.

Kripke, S. (1980) *Naming and Necessity* (Oxford: Blackwell, revised edition).

Losonsky, M. (1990) 'The Nature of Artefacts', *Philosophy* 65: 81–88.

Okasha, S. (2002) 'Darwinian Metaphysics: Species and the Question of Essentialism', *Synthese* 131(2): 191–213.

Owen, G.E.L. (1986). 'Aristotle on the Snares of Ontology', in G.E.L. Owen *Logic, Science, and Dialectic: Collected Paper in Greek Philosophy*: 259–278.

Putnam, H. (1973) 'Meaning and Reference', *Journal of Philosophy* 70: 699–711.

Quine, W.V.O. (1963) 'On What There Is', in W.V.O. Quine *From a Logical Point of View* (New York: Harper and Row): 1–19.

Rudder Baker, L. (2004) 'The Ontology of Artifacts', *Philosophical Explorations* 7: 99–112.

Rudder Baker, L. (2008) 'The Shrinking Difference Between Artifacts and Natural Objects', in Piotr Boltuc (ed.) *Newsletter on Philosophy and Computers, American Philosophical Association Newsletters* 7(2): 1–10.

Schaffer, J. (2003) 'Is There a Fundamental Level?', *Nôus* 37(3): 498–517.

Schaffer, J. (2009) 'On What Grounds What', in D. Chalmers, D. Manley and K. Wasserman (eds) *Metametaphysics* (Oxford: Oxford University Press): 347–383.

Schaffer, J. (2010) 'Monism: The Priority of the Whole', *Philosophical Review* 119(1): 31–76.

Schwartz, S. (1978) 'Putnam on Artifacts', *Philosophical Review* 87: 566–574.

Snowdon, P. (2009) 'On the Sortal Dependency of Individuation Thesis', in H. Dyke (ed.) *From Truth to Reality: New Essays in Logic and Metaphysics* (London: Routledge).

Snowdon, P. (1998/2009) 'Peter Frederick Strawson', in E.N. Zalta (ed.) *The Stanford Encyclopedia of Philosophy* available online at http://plato.stanford.edu/archives/fall2009/entries/strawson/

Strawson, P.F. (1974) *Subject and Predicate in Logic and Grammar* (Oxford: Oxford University Press).

van Inwagen, P. (1990) *Material Beings* (Ithaca: Cornell University Press).

Wiggins, D. (1979) 'Mereological Essentialism: Asymmetrical Dependence and the Nature of Continuants', in E. Sosa (ed.) *Essays on the Philosophy of Roderick Chisholm* (Amsterdam: Grazer Philosophische).

Wiggins, D. (1980) *Sameness and Substance* (Cambridge, MA: Harvard University Press).

Wiggins, D. (1995) 'Substance', in A.C. Grayling (ed.) *Philosophy 1: A Guide Through the Subject* (Oxford: Oxford University Press).

Wiggins, D. (2001) *Sameness and Substance Renewed* (Cambridge: Cambridge University Press).

Wiggins, D. (2012) 'Identity, Individuation and Substance', *European Journal of Philosophy* 20(1): 1–25.

Wilson, J. (1999) *Biological Individuality* (Cambridge: Cambridge University Press).

# 4    The Human Being Theory

The theory of individuation that has been set out in the preceding chapters finds distinctive expression in Wiggins's account of personal identity. In Chapter 7 of *S&SR* (a dramatically revised version of the one that appeared in *S&S*) he considers our identity judgments about persons and elaborates the procedures by which we trace *each other* through space and time.[1] What it takes for us to survive is a question that has a special importance for us in a way that questions about the persistence of chairs or of numbers do not. Whether we are faced with unusual (albeit true) stories of anterograde amnesia, or cases of extensive drug therapy or dementia, we routinely wonder what kinds of changes *we* can undergo. We are afraid of the things we cannot survive – and it is this concern, intimate and enduring, which nourishes the questions of the personal identity debate.

In accordance with the method outlined above, Wiggins's position is that we need to settle on the sortal concept under which we subsume ourselves before we can examine our principle of activity (and determine, from there, our persistence conditions). This leads to his Human Being Theory – a theory novel enough, and subtle enough, to invite multiple, sometimes contradictory, readings. Eric Olson interprets the Human Being Theory as a version of the Neo-Lockean account, wherein we are seen fundamentally to be persons – not animals – and our persistence conditions are those of psychological beings. In contrast, Harold Noonan and Peter Unger position Wiggins as an 'animalist', i.e. as holding that we are fundamentally human animals – not persons – with the concomitant biological mode of being. It is the aim of the first section of this chapter to show what these readings capture, and where they falter.

The problem is that neither of these interpretations engages with Wiggins's crucial claim that we are fundamentally *both* persons *and* biological beings. This is the thought that lies at the heart of the Human Being Theory – one that is simultaneously productive and controversial. Aside from the other worries it may excite, one is bound to wonder – as Snowdon[2] does – how we can be, *fundamentally*, more than one sort of thing. Wiggins himself notes that his proposal may appear to provoke the worries with relativity with which he started his inquiry.[3] To accept that we single ourselves out, correctly and fundamentally, as *both* is to accept the relativity of identity, anathema to the Leibnizian formulation of identity that Wiggins defends.

In the second half of the chapter, Wiggins's ingenious response to this concern is laid out. He claims that, though the terms 'person' and 'human being' differ in sense (just as '*equus caballus*' and 'horse' capture different aspects of that animal),[4] and though they may even differ in extension,[5] the concepts to which these terms allude are in some way *concordant*.[6] Our understanding of what a person is interweaves somehow with our understanding of what a human being is. The result of his argument is briefly stated in an opening passage from Chapter 7 of *S&SR*:

> [I]n so far as they assign any, the concepts *person* and *human being* assign the same underlying principle of individuation to A and to B and that that principle, the *human being* principle, is the one that we have to consult in order to move towards the determination of the truth or falsehood of the judgment that A is B.[7]

The *conceptual consilience* Wiggins divines between *person* and *human being* allows him to accommodate the insights of both Neo-Lockeanism and Animalism. It also provides a generative method for answering the puzzles of the personal identity debate. Assessing the identity conditions of persons leads into a well-trodden but nonetheless prickly thicket of philosophy; Wiggins's advice is to turn to the concordant concept *human being* to avoid it. Both concepts assign the same underlying principle of individuation – but in the latter case it is less obscure.[8]

Three central struts support the argument for conceptual consilience. The first is a Strawsonian point that our concept of a psychological being seems to allude to a biological substrate. The second relates to the *semantics* of the term 'person' – we use it, Wiggins argues, as though it were a *natural kind word*. The third strut is an argument from *interpretation and indexicality*, which develops out of Davidsonian thoughts about the conditions that must hold for us to be able to interact with one another in the way that we do. In each, Wiggins's focus is a putatively *pre-theoretical* concept, an element – like *substance* – of our conceptual scheme. My contention in the next chapter is that his *semantic analysis* misses its target, that our use of the term 'person' – inflected by cultural bias – is not a reliable basis for an examination of our (human) conceptual framework. By means of a genealogy, the object of his analysis is shown to be a cultural accretion, devoid of any unifying rationale – a *conception*, as Wiggins would have it, and not the *concept* itself.[9]

## 1  Neo-Lockean or animalist?

In Book II of the *Essay Concerning Human Understanding*, Locke defines a person as:

> a thinking intelligent being, that has reason and reflection, and can consider itself as itself, the same thinking thing, in different times and places.[10]

This definition is the linchpin of Locke's account of 'personal identity'. It describes that which Locke takes to constitute the identity of persons through space and time. It has been the subject of varied critiques but remains the lodestone for contemporary Anglophone discussions of personal identity.

Prefiguring Wiggins, Locke proceeds along sortalist lines and divines an intimate relation between understanding what sort a thing *is* and what it takes for such things to persist. For Locke we are fundamentally *persons*, that is thinking, *self-conscious* beings; so – he thinks – self-consciousness is a criterial property for personhood. Our survival, as self-conscious beings, depends on the continuation of our consciousness. Ultimately, for Locke, this continued consciousness is evidenced by the self-recording faculty of *experiential memory*. A memory of an earlier experience indicates that an individual is continuous with (identical to) the person that experienced it:

> [A]s far as this consciousness can be extended backwards to any past action or thought, so far reaches the identity of that person.[11]

In *S&S*, Wiggins's analysis emerges from his dialogue with Locke's. And crucially, he takes the Lockean account to disclose an important insight about the kinds of beings we fundamentally are. In *S&S* (and note well the footnote), he writes:

> There is something so interesting about the notion that a person is an object essentially aware of its progress and persistence through time, and peculiar among all other kinds of thing by virtue of the fact that its present being is always under the cognitive and affective influence of its experiential memory of what it was in the past; and this notion is so closely related, not only to profound contentions of Leibniz and Kant,[3] but also to deeply ingrained ordinary ideas of life as something to be reviewed and looked back upon; that I believe we should look with some suspicion at the contention that a continuity of consciousness condition of personal identity is irreducibly circular.

> 3 Cf. Leibniz, *Discours de Metaphysique* XXXIV...; J. Bennett, *Kant's Analytic*, Cambridge 1963, p. 117, 'the notion of oneself is necessarily that of the possessor of a history: I can judge that this is how it is with me now, only if I can also judge that that is how it was with me then. Self-consciousness can coexist with amnesia – but there could not be a self-conscious person suffering from perpetually renewed amnesia such that he could at no time make judgments about how he was at an earlier time.[12]

This thought about the importance of experiential memory is restated and modified in *S&SR*:

> [W]hat I am in the present ('my present self') always lies under the cognitive and affective influence of what I remember having been or having done

or undergone in the past, no less than of that which I intend or am striving to make real in the present or the future. But if it is the nature of persons to be remembering beings whose conception of what they themselves are is all of a piece with their experiential memory, then some constitutive connexion ought to be expected (it will be said) between their experiential memory and their identity.[13]

In the first quotation Wiggins suggests, along the Kantian lines contained within the footnote, that we cannot help but experience ourselves as remembering beings. It is part of our conceptual framework that earlier thoughts underpin our present ones. As the second quotation makes clear, this thought disappears in later work. But he continues to view personal memory as fundamental to our nature. Add to this his overt description of his approach as 'neo-Lockean' (in, for example, 'The Person as Object of Science')[14] and it is unsurprising to find interpretations of his work as endorsing some kind of psychological persistence condition for persons. It is for these reasons that Olson writes, in *The Human Animal*:

> Whereas my view is that psychological continuity is completely irrele-vant, except derivatively, to our persistence, Wiggins insists that certain broadly mental capacities – sentience, desire, belief, motion, memory and others – are part of what it takes for a person to remain alive, and so to continue existing.... Wiggins argues [that memory] is '*crucially relevant to our choice of continuity principle for determining the biographies of persons*'.... Although there is much in Wiggins's work that I do not understand, his view seems to me to be a sophisticated version of the Psychological Approach.[15]

My contention, however, is that this is a misreading of Wiggins's position. Wiggins does not take experiential memory to indicate anything about identity. He is sympathetic to Locke's position (and in his earlier work more so), but he never goes so far as to endorse an exclusively psychological criterion of identity. He rejects the view that our survival stands or falls with continued conscious-ness. Memory is not called on as evidence for survival.

It is understandable that Olson – whose focus is on *S&S*, and an earlier paper, 'Locke, Butler and the Stream of Consciousness'[16] – reads Wiggins as a sup-porter of Locke. In both of these texts, Wiggins carefully defends the Lockean thesis against various critical descendants of the objections raised by Joseph Butler in *First Dissertation*.[17] Yet, even though Wiggins denies any circularity or absurdity in Locke's consciousness criterion,[18] he also denies that memory can provide any significant basis for identity judgments and 'is doomed always to bring too little too late' to the analysis.[19] (And though, of course, its publica-tion followed Olson's *The Human Animal* by four years, *S&SR* explicitly dis-tances itself from that earlier position and recants any dubiety about Butler's critique.[20])

Wiggins finds in Butler an insurmountable obstacle to the Lockean account. The famous objection, to which *S&SR* pledges full support, runs thus:

> One should really think it self-evident, that consciousness of personal identity presupposes, and therefore cannot constitute, personal identity, any more than knowledge, in any other case, can constitute truth, which it presupposes.[21]

The Butlerian thought, which appears obscurely in *S&S* and fully in *S&SR*, is that determination of a *genuine* memory invokes *another* account of identity and so cannot *constitute* it.[22] In *S&SR* (shorn of the previous arguments against absurdity and circularity) the point is demonstrated with an example case of putative remembering. Imagine, Wiggins writes, that we know that A once inadvertently caused a fire in the Chigwell College of Commerce. Suppose that B seems to remember doing this. That *appears* to suggest that B is A. However, Wiggins continues,

> B is only the same person as A if his seeming to remember is his really and truly remembering setting fire to the book stack.[23]

It must be a genuine memory.[24] But how can we be sure that it *is* genuine? How can it be established that B is not, for example, subject to some bizarre hallucination or suffering delusions? Here we reach the nub of Butler's point, as Wiggins reads it:

> Where someone appears to remember starting that fire, they can't be right unless they were indeed there at the fire.[25]

Which is to say *another account* of identity must be invoked to explain what it is for someone to do something and then *genuinely* remember doing it. In section 4 we will see how this thought segues into a positive thesis about our spatio-temporal continuity being understood by reference to some material foundation – and how Wiggins subsequently claims that this material foundation can only be an animal, specifically a *human being*. The intermediary point is simply that Olson's characterization of Wiggins as endorsing an essentially 'psychological' criterion of identity does not, and cannot, correspond to Wiggins's Butlerian critique of Locke.

## 2  Quasi-memory

Before turning to the material foundation, and the Human Being Theory, it is necessary to note another avenue that this discussion of Butler opens up, and one which receives considerable attention in *S&SR* and the correlative secondary literature: the issue of *quasi-memory*.

As Wiggins points out, the most obvious way of distinguishing between real and apparent memories is to say that the genuine memory – and not the delusion – is of an experience the rememberer herself actually *had*; as was evident to Butler, this creates a circularity in Locke's account. Yet Sydney Shoemaker and Derek Parfit have argued that genuine memories and delusions can be distinguished by yet other means: it is not, they say, about *who had* the original experience, but rather about how the subsequent memory-experience was *caused* that determines whether or not a memory is delusional.[26] Thus, Napoleon's memory-like experiences of Waterloo are not delusional because they are causally connected to the events at Waterloo in the right way, while the same is not true of the memory-like experiences of George IV (who was never there, but believed he was).[27] So, Parfit writes:

> To answer this objection, we can define a wider concept, *quasi-memory*. I have an accurate quasi-memory of a past experience if (1) I seem to remember having an experience, (2) *someone* did have this experience and (3) my apparent memory is causally dependent, in the right way, on the past experience.[28]

It is along these lines that some neo-Lockeans deny that memory-experiences presuppose identity (citing cases of memory-transplantation to show how an individual can have a non-delusional quasi-memory of someone else's experience).[29] By this means they hope to circumvent Butler's circularity objection and any reformulations by those like Wiggins. Consequently, a considerable portion of Chapter 7 in *S&SR* is devoted to replying to this neo-Lockean line.

While my aim here is only to acknowledge the discussion, it is helpful to briefly consider Wiggins's response. Pointing to Parfit's definition of quasi-memory he asks whether condition (3) can admit *incomplete* or *imperfect* or *partially wrong* or *oddly produced* memories as quasi-memories. (For these are things we surely have: you may remember your last birthday party, perhaps, but what clothes were you wearing and who was there?) As Wiggins puts it, trenchantly:

> The thing we see that Parfit presents [in the passage above] is not a definition of 'quasi-remember' or 'quasi-memory' at all. It's a definition (he himself announces that it is a definition) of 'have an accurate quasi-memory'. Inaccurate quasi-memory is not provided for.[30]

It's true that the neo-Lockean may dispute Wiggins's analysis (Shoemaker, certainly, seems to).[31] In any case, quasi-memory is not the focus of the present work and more paper will not be added to the reams already spent on it. There are, as we will see, other, more interesting, disagreements between Wiggins and the neo-Lockeans.

## 3  Against an 'animalist' reading

The focus being put on this *material/biological foundation*, combined with the critique of Lockeanism (and neo-Lockeanism), suggests that Wiggins will ultimately endorse some variant of an 'animalist' thesis. In Olson's dichotomous terms, he appears to propound a 'biological' theory, while rejecting a 'psychological' one.[32] This is how Harold Noonan[33] and Peter Unger[34] both read him – i.e. as claiming that we are fundamentally human animals, and *not* fundamentally persons, and that our persistence conditions are those of the animals that we are. Yet this is another misreading, once again capturing some of the story, but not all.

Wiggins certainly holds that we are, fundamentally, a kind of animal (members of the species *homo sapiens*, to be precise). However, as noted, he is also deeply impressed by the Lockean thought that we are conscious, remembering beings. The kind of picture outlined by the animalist, which denies that we are fundamentally persons, does

> insufficient justice to a line of reflection still prompted by John Locke's account of these things: what I am in the present ('my present self') always lies under the cognitive and affective influence of what I remember having been or having done or undergone in the past, no less than of that which I intend or am striving to make real in the present or the future.[35]

Wiggins denies that continued consciousness is a condition for our survival (according to the Human Being Theory we may survive in vegetative states) – but we are beings who typically have a rich psychology, who have potential for *Bildung* (as will be discussed),[36] who are, among other things, *rememberers*. The animalist account fails to capture these complexities and Wiggins has little or no rapport with it.

A terminological point can be entered here. Despite the (relative) popularity of the term 'human animal', following the publication of Olson's book in 1997, it does not appear in *S&SR* or in Wiggins's subsequent work.[37] Instead, Wiggins talks of 'human beings'. The interpretation I have given above offers an explanation. The two terms differ in sense: the first refers to what, following David Bakhurst, we may call a 'mere animal'; the second, to biological beings that, as John McDowell puts it, are 'at home in the space of reasons'.[38] 'Human being' does not restrict the area of inquiry to the biological (the ambit of 'biological' is discussed in Chapter 6).[39]

Wiggins is not a neo-Lockean (*pace* Olson), but nor is he an animalist (*pace* Noonan and Unger). He agrees with the animalists that our persistence conditions are those of the organic beings that we are, but he would object to the rejection of the neo-Lockean claim that we are fundamentally persons. At the same time, he rejects the neo-Lockeans' exclusive focus on psychological continuity, which overlooks the biological aspects of our nature. He holds instead that we are fundamentally human beings *and* persons. This is a claim he defends by focusing on the *consilience* of those concepts – a connection explored in the next section.

## 4 Conceptual consilience: three arguments

For Wiggins, our understanding of what a person is interweaves, somehow, with our understanding of what a human being is. This is a thought that he tries to capture in the appended diagram (Figure 4.1), found in 'Person as the object of science' (with due deference to Frege).[40]

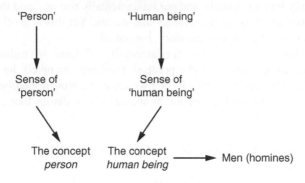

*Figure 4.1*

Though the terms 'person' and 'human being' differ in sense, *person* and *human being* refer (in the same way that *horse* and *Equus caballus* do) to 'the same things out there in nature' ('homines').[41] Wiggins presents different, inter-linking arguments to substantiate this connection.[42] He presents a *Strawsonian argument* to the conclusion that persons must be *material* things. There is also a *semantic argument*, grounded in the way we use 'person' as though it were a natural kind word. And lastly, but perhaps most importantly, there is an argu-ment from *interpretation and indexicality*, which collects together various insights from Donald Davidson.

### (i)  The **Strawsonian argument**: *a preliminary link*

It is Peter Strawson's thought that *person* is a 'primitive concept' that forms the initial connecting cord between that concept and *human being*.[43] Wiggins uses Strawson's analysis to augment the Butlerian critique (presented above) to claim that our concept of a thinking being must also be of a material being. While there is some disagreement about what exactly Strawson means by 'primitive',[44] Wig-gins's interpretation (followed here) runs thus:

> [A] person is, *par excellence* (and as a presupposition of all the traditional questions in the philosophy of mind), the bearer of *both* M-predicates *and* P-predicates, where M-predicates are predicates that we could also ascribe to material objects and P-predicates are predicates that we could not pos-sibly ascribe to material objects and comprise such things as actions, inten-tions, thoughts, feelings, perceptions, memories, and sensations: and that 'a person' is a type of entity such that both predicates ascribing states of con-sciousness and predicates ascribing corporeal characteristics are equally applicable to an individual of that single type.[45]

Strawson's claim develops out of his descriptivist critique of Cartesian mind/body dualism. His thought is that the idea of an immaterial, thinking thing collides with a basic principle about psychological thought, made evident by our practices of thought ascription. The idea is that one can only ascribe experiences to oneself if one is prepared to attribute them to others. And to do this, Strawson claims, one must be able to fix on other subjects. They cannot be – *contra* Descartes – *non-spatial*.[46]

What is significant for present purposes is a particular objection that Wiggins levels at this Strawsonian account, to which he is otherwise largely sympathetic. Strawson notes that, while our practices of self-reference require that we be bearers of P-properties and M-properties, we can *conceive* of ourselves as lacking P-properties (for example, as comatose or unconscious individuals).[47] Yet, in the same way as the material body can survive the loss of psychological properties, so too – says Strawson – we can conceive of a person's consciousness outliving her body:

> [E]ach of us can quite intelligibly conceive of his or her individual survival of bodily death. The effort of imagination is not even great.[48]

While Wiggins agrees that we can conceive of ourselves as lacking psychological properties, as Snowdon points out,[49] he takes issue with this overly even-handed treatment of the mental and physical aspects of a person. He suggests that the situation Strawson describes – of a consciousness outliving her body[50] – clashes with how we ordinarily conceive of psychological experiences. The decisive proposal in his 1987 paper 'Person as the Object of Science' is that, once it has been sufficiently worked out, we will find that the notion of a bearer of P-properties will *necessarily* involve ascription of M-properties. That is to say that psychological states and capacities are, in some way, essentially 'matter-involving'.[51]

Wiggins does not aspire to prove the confluence of P-properties with M-properties in its full generality.[52] But in the 1987 paper he attempts to demonstrate its plausibility by presenting studies of psychological events that cannot but be conceived of as involving something material. Of prime interest is his conceptual analysis of *remembering*, in which we find the crucial intersection of his thoughts about Strawson and the earlier, Butlerian critique of the Lockean memory criterion.

Consider again the example given above: A's setting fire to the College of Commerce and B's memory of causing the fire. How do we understand the claim that B *remembers* this? Not, surely, as meaning only that B has some kind of agent-centred inner representation of the event. A delusional might have such a representation too. The point is that we must also think there is the right sort of *causal* relation between the act and the subsequent memory-experience. Something else must be invoked. And it is at this point that Wiggins turns to an influential – but now (ironically) forgotten – paper, by C.B. Martin and Max Deutscher: 'Remembering'.[53] In 'Remembering', Martin and Deutscher essay a claim about what exactly it is to be the 'right sort of causal relation', a claim that Wiggins characterizes in the following way:

[I]t is impossible to say what the right sort of causal connection between an incident and memory representation of it is without having recourse to the notion of something like a memory trace.[11]

11  The memory trace may be conceived under the specification 'the normal neurophysiological connection whatever that is, between rememberings and the incidents of which they are
rememberings'... Deutscher and Martin carefully explore a multiplicity of alternatives to
the explicit memory-trace account of the causal connection between incident and experiential memory of incident. They show that none of these accounts can simultaneously allow
for the possibility of prompting and define the particular sort of operativeness we are
looking for between incident and representation.[54]

According to this analysis of remembering, we cannot conceive of memory causality as 'a transaction over a matterless gap between the external world at one
time and a mind at a later time'.[55] Thoughts about remembering and memory
necessarily involve (overtly or otherwise) some conception of the normal sort of
*bodily* process that issues in the inner representation of a past experience – i.e.
we cannot conceive of an individual actually remembering something without
also thinking that their memory is the result of a material process, which reaches
back to the initial activity.[56,57]

How stable is Wiggins's position here? Snowdon identifies one potential difficulty:[58] while it may be the case that remembering and perceiving and other
such psychological states require some *material* foundation, why, he wonders,
must this foundation be organic, 'biological' and 'living'? That, surely, is what
Wiggins needs if he is to tie the knot between the relevant concepts: and it is a
live issue whether or not psychological states can only be had by organic beings.
(And certainly, science fiction stories furnish us with numerous examples of
robotic intelligences, as well as immaterial consciousnesses.) Of beliefs and
desires, alongside other psychological states, Snowdon writes:

[There is] nothing obviously biological in the idea of these structures.[59]

Closer attention to Wiggins's texts reveals an implicit response. Consider his
analysis of perception, which also appears in the 1987 paper:

For there to be a perception of *x*, something would have to be able to count
as a misperception of it. But what is the difference? If we are to make the
distinction we need, then there has to be something independent of what is
subjectively given in perception. But then we must ask the position of the
perceiver. There must be such a thing as an answer to the question of where
the perception is *from*. Otherwise there is nothing that the perception is
answerable to. And what else can fix where the perception is from but the
body, head, and eyes of the perceiver?[60]

This passage indicates the importance of the *assemblage* of the material foundation (in a way that the discussion of memory does in more abstract terms).

'Perception' is clearly meant to be visual perception, and visual perception requires some suitable arrangement of the body, head and eyes. The predictable response to this will be that a suitably structured robot (i.e. one with cameras and some kind of recording unit) might be said to 'perceive'. (Here though, as with computer 'memory', there is the possibility of this sliding into metaphor.[61]) But consider also – Wiggins might say – the *desires*, experienced by persons, to which Snowdon denies any 'biological' element.[62] Certain desires surely seem to contradict this. After all, in the same paper, Snowdon himself points to our over-whelming sense of being of a certain sex.[63] If we include sexual desire among the states of persons, particular kinds of biological structure *do* seem to be neces-sary. Thinking of persons as entities enjoying a variety of psychological experi-ences, including sexual desire, we shall find it harder and harder not to think of them as having a particular kind of *biological* makeup.[64]

Maybe it will be said that – irrespective of love, desire, hatred (and other such arguably paradigmatic psychological states) – the crucial question is whether or not our idea of *memory* involves conceiving the *rememberer* as a biological being. And this perhaps is less obvious than it is in the case of sexual desire. Wiggins's first recourse is to Martin and Deutscher's discussion of remembering and to the centrality of the notion of a 'memory trace'. If one is persuaded that our ideas about memory necessarily involve the notion of this 'memory trace', one might also be persuaded that rememberers can only be beings that possess those special capacities afforded by our neurophysiological make-up.

I do not think Wiggins's position stands or falls with Martin and Deut-scher's analysis. It is consonant with his general approach to emphasize instead the complexity and the deeply integrated nature of our psychological states. This harks back to a point made in Chapters 1, 2 and 3: that Wiggins opposes the drive towards a *piecemeal* approach to philosophical issues and the determination to separate areas of study into smaller, more 'easily digesti-ble' chunks, without attending to the connections between them. On balance I think the best option for Wiggins is to say that a memory is not a distinct, iso-lable object of study. Memory is all of a piece with auditory and visual and emotional states which are more clearly reliant on a biological nature. He may respond to the kind of question proposed by Snowdon by asking how separ-able psychological states (desiring, believing, imagining, remembering...) actually are, and whether we can conceive of memory as devoid of any other aspect. If we focus on these interrelations, and how some psychological states must be borne by biological beings, then we shall find it hard to divorce per-sonhood and animalhood (if not yet human beinghood). Such are the thoughts contained within the passage from Hegel that Wiggins quotes at the start of his chapter on personal identity in *S&S*:

It is only in its proper body that mind is revealed. The [idea of the] migra-tion of souls is a false abstraction, and the physiology ought to have made it one of its axioms that life had necessarily in its evolution to attain to the human shape as the sole sensuous phenomenon that is adequate to mind.[65]

The thought that our psychological states can only be enjoyed by beings with a particular physiology – *human beings* – is a speculative connection and not one that Wiggins tries to strengthen (the quotation from Hegel does not appear in the later texts).[66] He takes it to be suggestive but not conclusive; the argumentative work is done by his *semantic argument* and his argument from *interpretation and indexicality*.

### *(ii)  The* semantic argument *and the animal attribute view*

As a descriptivist, Wiggins is interested in elucidating the structure of our conceptual framework. And in *S&S*, and to a certain extent *S&SR*, his method for doing so involves, in no small part, an examination of the things that we find ourselves saying about *persons*.[67] Following his paper 'Locke, Butler and the Stream of Consciousness', a central aim of *S&S* is to present an explicitly 'descriptive' analysis of 'person'.[68] He investigates how we commonly apply the term in order to elucidate its meaning and the conceptual requirement it fulfils.[69]

Wiggins takes Locke's account of 'person' as his point of departure – and while his sympathies with Locke have fluctuated, his objection to the Lockean definition of 'person' has remained unwavering. He describes Locke's definition as an 'analytical excogitation of a nominal essence'.[70] That is to say that in the *Essay*, 'person' is not taken to introduce a real essence in the way that a natural kind term would do; it *stipulates* one, as do the terms we use to refer to artefacts and positions of authority.[71] A central strand of Wiggins's work has been to argue that this rendering of 'person' clashes with our use of that term in everyday practice. We do not use it as though personhood is a role to be fulfilled or as a functional specification. Thus, Wiggins writes (and mark well the descriptivist entreaties to 'our innermost convictions' etc.):

> Nobody thinks of the persons we actually encounter in nature as artefacts, or as having identities which are 'for decision' as artefact identities are sometimes for 'decision' when there is a changing of parts.[72]

> A pure conventionalist view of the identity of people would fly in the face of the innermost convictions of almost everyone.[73]

> [I]t flies in the face of the innermost convictions of almost everyone to try to think of the persons we encounter in nature as having identities that are any less determinate than the identities of animals.[74]

> The definition of *person* is not something we conceive for ourselves in the way in which we have conceived for ourselves the nominal essences for *hoe* or *house*.[75]

'Person' (Wiggins claims) functions in our everyday practice as though it were a *natural kind word*.[76] We do not stipulate, as the constructionists do (discussed below), what 'person' means. Rather we learn more about persons by singling them out *in rerum natura*.[77] It is not by consulting textbooks or dictionaries that we learn what persons are, but by encountering them in the world. And when we try to adumbrate the marks of personhood – consciousness, memory, imagination, love, intelligence ... – we find an essential *aposiopesis*,[78] spaces we have yet to fill. The attributes of persons are not circumscribed by a definition; we constantly add to them as we discover more about ourselves and each other.

Yet even while Wiggins takes these features as indicators that 'person' is *akin* to a natural kind word, he is cautious of giving it the full status of such terms as *rabbit, ivy, dog* and (putatively) *human being*.[79] In *S&S* at least, he suggests that we see 'person' as something like a *qualification* of a natural kind determinable, a 'hybrid concept' with a natural kind element and a systemic element as well[80] (compare the way 'vegetable' collects together a group of savory, edible plant kinds).[81]

Whether it functions *as* a natural kind word or as *akin* to one, this descriptive analysis appears to indicate that the *person* concept intersects with some idea of a natural, animal kind. This is something Wiggins takes to be clear in our use of that term. It thus explains why we feel the strain we do when trying to understand whether robots are persons – since, we may think, they have shallow nominal essences[82] – and when we think of ourselves as 'artefact-like' – or as subject to transplantation or to arbitrary replacement of parts.[83]

> [I]t is certain that we still believe that, to have genuine feeling or purposes or concerns, a thing must *at least* be an animal of some sort.[84]

Wiggins holds that the richness of our use of 'person' suggests that it has to be allied in some way with a natural kind concept. In *S&S* this line of argument culminates in what Wiggins calls the 'Animal Attribute View'.[85] He sets it out as follows (note the dots which show how this elucidation of person is not a strict definition, but an essentially incomplete statement, to be filled out by looking to persons as we encounter them):

> Perhaps *x* is a person if and only if *x* is an animal falling under the extension of a kind who typical members perceive, feel, remember, imagine, desire, make projects, move themselves at will, speak, carry out projects, acquire a character as they age, are happy or miserable, are susceptible to concern for members of their own or like species ... [note carefully these and subsequent dots], conceive of themselves as perceiving, feeling, remembering, imagining, desiring, making projects, speaking ... have, and conceive of themselves as having, a past accessible in experience-memory and a future accessible in intention.[86]

The frailties of this treatment are considered in depth in the next chapter, but it is worth noting here that, however effective this argument is, it is not enough to substantiate a specific connection between *person* and *human being*. Even if this semantic analysis of 'person' appears persuasive, the connection it forges is between *person* and *animal*. There is, as yet, no special association with the concept *human being*. As Wiggins himself notes:

> [I]t is not absolutely excluded ... that the extension of *person* should give hospitality to such creatures as chimpanzees or dolphins or even, in exchange for suitably amazing behaviour, to a parrot. According to this view, a person is any animal that is such by its kind as to have the biological capacity to enjoy fully the psychological attributes enumerated; and whether or not a given animal kind qualifies is left to be a strictly empirical matter.[87]

More needs to be said. The extra link is provided by another argument, which appears alongside the semantic one and works in concert with it. This argument has come to increasing prominence in Wiggins's later texts. This is an argument from *interpretation and indexicality*.

### *(iii) An argument from* interpretation and indexicality

'Interpretation' here is a term of art and requires some unpacking. It refers to a discussion found in the work of Donald Davidson – another figure who, alongside Strawson, exerts a keen influence in Wiggins's philosophy.[88] In his paper 'Radical Interpretation',[89] Davidson sets out to explain what exactly it takes for us to be able to understand one another – what it is that makes our interpretation of the linguistic behaviour of a speaker possible.[90]

Imagine yourself confronted by someone from a vastly different socio-cultural background from yours, one with whom you do not share a language. What happens in this encounter – how can you begin to understand, to interpret, this stranger? In broom-broad brush-strokes, Davidson sees there to be two important assumptions on the part of the interpreter (assumptions which he calls, collectively, the 'principle of charity').[91] The first is that the speaker's behaviour satisfies strong normative constraints – i.e. that she reasons in accordance with logical laws. The second is that our interpretive procedures depend on our thinking the speaker normally says what she believes to be true, and that something's being true. In order to make this second assumption, the interpreter must overcome innumerable unknowns (for how can one know what someone else believes?), and doing this, Davidson suggests, she *projects* herself into the subject's position, and assumes that the speaker believes or would believe what she, the interpreter, would believe in her position. That is, she must see the speaker *as a being like her*.[92] She sees the speaker as a being with whom she can get onto 'the same wavelength'.

How does this connect with Wiggins's work? Wiggins is interested in our everyday practices, and Davidson's discussion gives an insight into certain everyday, linguistic procedures. These interpretive assumptions are the ones we have to make in order to interact with each other in the way we do. For interpretation to be possible, one has to see others as being like oneself. We must see each other as 'subjects of interpretation',[93] beings with whom we may get onto 'the same wavelength' or 'on net'.[94] Crucially, Wiggins sees these requirements as *underwriting* our use of the term 'person'. In *S&SR* he writes:

> Par excellence, a person is ... a subject ... of interpretation, a being that both interprets and is interpreted.[95]

And, more recently, in 'Identity, Individuation and Substance':

> [A] person is ... a creature that interprets other human creatures and is wide open to be interpreted by them.[96]

Understanding *person* in this way makes good sense of the *aposiopesis*, typical of natural kind terms, which Wiggins identifies in our usage. Moreover, understanding persons as 'subjects of interpretation' offers a principle[97] by which, in the absence of analytical determination, the marks of personhood may be enumerated:

> No wonder ... that, in the interest of our securing and vindicating our mutual attunement, the Lockean elucidation of 'person' grows and grows. For there is no clear limit to what concerns and capacities and perceptions and feelings ... we shall have to credit our fellows with if we are to make sense of them.[98]

We cannot give a definitive list of the marks of personhood because the concerns and capacities we must be attuned to in order to interpret are not static or restricted. Wiggins emphasizes the procedural richness of putting oneself 'in another's shoes' in *S&S*.[99] The interpreter's imaginative act cannot just be about what the speaker is seeing, hearing, etc. It must also be about recognizing the kinds of *concerns* they will have and e.g. the kinds of objects that will be of interest to them, and why.[100] For interpretation to take place the interpreter needs to 'know more than nothing not only about the world but also about men in general'.[101]

Finally it begins to emerge how, in addition to supporting a link between *person* and some animal kind, interpretation secures an important connection between *person* and *human being* – because 'person' is understood to be implicitly *indexical*. Persons are creatures *like us* (with whom we may get 'on net'), where we are *human beings*.

> [W]e shall only count something as a person if it is the kind of thing that we who are human beings can interpret and can make sense of in a manner that is in principle not delimited or circumscribed.... People are creatures of a kind to be the subjects of fine-grained interpretation *by* us, who are human beings, and to be the putative exponents of fine-grained interpretation *of* us.... The thing we are concerned with here is a rationality of ends (as well as of means and the fit of means to ends). For these purposes, our only

usable paradigm or stereotype of a reasonable being, or of a rational conscious being whom we can interpret, or of a person, is that of a human being. Our only proxy for a thinking, feeling soul is a striving, symbol using/misusing, embodied human being.[102]

Wiggins disclaims any aspirations to give a 'transcendental argument' for the conceptual connection between *person* and *human being*.[103] But the aim of the arguments set out above is to tease out why we may think a conception of a non-human person stretches the concept too far. It is not *logically* excluded that other animals could be persons. But see, he says

> whether you can describe, up to any required level of detail, how we should make sense of non-human creatures, become attuned to them, or be in a position to treat their feelings as if they were our own; or see how you imagine making sense of these creatures *without* doing that'.[104]

It is hard to think we could understand Martians or dolphins or automata as anything other than inscrutable alien intelligences.[105] To refer, as Wiggins does, to one of the most-cited philosophical summaries of this view:

> If a lion could speak, we could not understand him.[106]

Moreover, in reading 'person' in conjunction with these Davidsonian insights about interpretation, Wiggins can show exactly *why* the Lockeans think as they do. Experiential memory is so central to our view of ourselves because it plays a central role in interpretation. When people do or suffer something, this impresses itself on their mind, colours their experience and influences their future responses.[107] Nevertheless, Wiggins resists too strong a Lockean impulse. As he puts it:

> Locke can be right about remembering as central among the marks of personhood without Locke's or the neo-Lockeans' being right about personal identity.[108]

*   *   *

These arguments suggest a link between the concept *person* and the concept *human being*. Insofar as they are successful they allow Wiggins to collect together the attractive elements of both neo-Lockeanism and Animalism. Confused by the intricacies of *person* we can, he says, turn to the concordant sortal concept *human being* when answering the questions of the personal identity debate.[109] That substance sortal encapsulates a *principle of activity* or 'mode of being' for members of that sort – a principle from which persistence conditions may be extrapolated and questions about survival answered.

According to Wiggins, the 'human being principle' is the theoretical description that we find (via the deictic-nomological method described in Chapter 3) holding between exemplars of that kind. It includes all the biological processes characteristic of human beings: the way we typically breathe, walk, digest, reproduce, grow, age and die. It also includes typical behaviour, social tendencies and manner of interaction. Each human being affords this principle specific determination; it is this specialized and refined determination of the human mode of being which we track when we individuate ourselves.

The Human Being Theory thus suggests determinate answers to a number of classic puzzle cases. The amnesiac is seen to be identical with the person who suffered the event that resulted in the amnesia. Memory does not register on identity.[110] The same is true for the comatose patient. She is identical with the person who suffered the accident that caused the coma. Even though she has lost consciousness (never perhaps to regain it) she continues to be the kind of entity that has those characteristic and uncircumscribed capacities.[111] Wiggins does not mention patients in vegetative states,[112] but he can plausibly say they continue to realize a particular principle of activity.[113] The foetus, too, is identical with the adult it will later become, realizing, as it does, that same principle of activity.[114] (Mention will be made of the brain transplantation story, and Wiggins's response to it, in Chapter 9.)

Wiggins's engagement with the personal identity debate is subtler than most. Gathering together the insights of Strawson and Davidson, he elucidates the meaning of our everyday concept of *person* and the thoughts that underpin our use of the word. With these elucidations now elucidated, questions will be raised about the real strength of the lien between *person* and *human being*.

## Notes

1  Wiggins 2001: xiii.
2  Snowdon 1996: 35.
3  Wiggins 1987:

> Well, someone may say, whatever truths there may be to discover about this, the expression 'a person' obviously doesn't mean the same as the expression 'a human being'; and neither means the same as 'a self': so being a person isn't the same as being a human being or being a self. These are different concepts. And so, he may say, moving straight up to the higher or more transcendent ground 'x can be the same person as y without being the same human being as y, and x can be the same self as y without being the same human being as y'.
>
> (57)

4  Wiggins 1987: 59.
5  Wiggins 2001: 193.
6  Wiggins 1996: 248.
7  Wiggins 2001: 194.
8  Wiggins 2001: 193–194.
9  Wiggins 2001: 10. What is the difference between a concept and a conception? Snowdon gives a helpful overview in his paper 'On the Sortal Dependency of Individuation Thesis':

> In [Wiggins's] sense of concept, no one possesses them, and no one lacks them; they are not acquired or applied; and no mind is necessary for them to be there. Wiggins

calls his use of "concept" *Fregean*. The cognitive correlate of a concept Wiggins calls a conception. That is, so to speak, a mental focusing on, or perhaps a means of mentally focusing on, a concept. Conceptions are said to be *Kantian* concepts.

(Snowdon 2009: 256)

10  Locke 1690/1975: II, xxvii, 9.
11  Locke 1690/1975: II, xvvii, 9.
12  Wiggins 1980: 150–151.
13  Wiggins 2001: 196.
14  Wiggins 1987: 68.
15  Olson 1997: 20–21.
16  Wiggins 1976.
17  Butler 1736.
18  E.g. Wiggins 1976: 132–136.
19  Wiggins 1976: 142.
20  Wiggins 2001:

the new chapter on personal identity, focuses on human beinghood, and recants anything I have ever said against Bishop Butler's objection to Locke's account of personal identity.

(xiii, see also 204, fn. 12)

This shift helps to account for his positioning of himself as a 'neo-Lockean' in the earlier work, but not the later.

21  Butler 1736.
22  Cf. Wiggins 1976: 142f, 1980: 161ff, and 2001: 203ff.
23  Wiggins 2001: 204.
24  NB This worry appears in Wiggins 1976, and 1980, as the concern that 'C* offers no account of error'. E.g. Wiggins 1976: 138.
25  Wiggins 2001: 204 (original italics).
26  Parfit 1984: 220 (see also Shoemaker 1970: 269–285).
27  Wiggins 2001: 215.
28  Parfit 1984: 220.
29  Parfit gives the example of Jane, who agrees to have Paul's memory-traces implanted in her brain (1984: 221) These kinds of science fiction thought experiments are critiqued in Chapter 9.
30  Wiggins 2001: 224.
31  See Wiggins's and Shoemaker's discussion in *The Monist* (Shoemaker 2004a, Wiggins 2004a, Shoemaker 2004b, Wiggins 2004b).
32  Olson 1997: 7ff.
33  Noonan 1998: 302.
34  Unger 1990: 120–123.
35  Wiggins 2001: 196. To my knowledge, Wiggins nowhere explicitly talks about 'Animalism'. This passage appears as part of a defence of the Human Being Theory, in which he denies that it neglects the Lockean concern with the psychological (as Animalism, as construed, also does).
36  'Bildung' refers to a form of 'self-cultivation', a process of personal and cultural maturation.
37  Others use the term frequently to describe his view – even those who do not appear to position him as an animalist (e.g. Bakhurst 2005).
38  McDowell 1994: 125. Bakhurst brings McDowell to bear (in Bakhurst 2005).
39  There is another possible reason why Wiggins will prefer 'human being' to 'human animal'. 'Human being' has a different linguistic pedigree from 'human animal' – where the latter is a technical term that exists almost exclusively within philosophical discourse, the former is not.

40  Wiggins 1987: 60.
41  Wiggins 1987: 60.
42  These arguments appear in various forms in Wiggins 1976, 1980, 1987, 1996, 2001, 2005, 2012.
43  Strawson 1959: Chapter 3.
44  E.g. Ishiguro 1980.
45  Wiggins 1987: 63–64.
46  Strawson 1959: Chapter 3. See also Snowdon 1998/2009 for a helpful commentary.
47  There is no tension here because he sees this idea, of our lacking psychological properties, as 'secondary' to the primary concept of person (Strawson 1959: 115).
48  Strawson 1959: 115.
49  Snowdon 1996: 33.
50  Admittedly, Strawson describes how quickly such a ghost-like entity would fade away (1959: 116) – but the point still stands.
51  Wiggins 1987: 64 (see also Snowdon 1996: 33).
52  Wiggins 1987: 64.
53  Deutscher and Martin 1966.
54  Wiggins 1987: 65 and 209, fn. 11.
55  Wiggins 1987: 65 and 209, fn. 12.
56  What of 'computer memory'? It is likely that Wiggins will dismiss this as metaphorical usage, secondary, if not tertiary, to our everyday conception of memory.
57  The argument in the 1987 paper thus stands as a structural development in Wiggins's work. In 1976 and 1980, he registers Butler's concern with the Lockean criterion, that the memory condition cannot be sufficient, because it presupposes another account of identity. His 1987 paper marks an explicit transition from this critique to the positive thesis that this alternative account involves something *material* and, as indicated, *physiological*.
58  Snowdon 1996: 44.
59  Snowdon 1996: 44.
60  Wiggins 1987: 65.
61  Snowdon 1996: 40.
62  Snowdon 1996: 44.
63  Snowdon 1996: 35.
64  This does not mean we cannot (perhaps) *synthesize* persons – but we are committed to thinking that to synthesize them we must effectively synthesize the workings of the human body. Wiggins notes this repeatedly, e.g. 1996: 247, and 2001: 90.
65  Wiggins 1980: 148, quoting Hegel, *Lectures on Fine Art*.
66  The thought is preserved, however, in Wiggins 2001: 198–199.
67  Wiggins 2001: 2.
68  Wiggins 1980: 149.
69  See also Wiggins 1976: 149, 151, 1987: 62–63.
70  Wiggins 1976: 151.
71  Wiggins 1987: 63.
72  Wiggins 1976: 151.
73  Wiggins 1976: 151.
74  Wiggins 1980: 171.
75  Wiggins 1980: 173.
76  Wiggins 1980: 171, 1987: 62, 2001: 238f.
77  Wiggins 1987: 62.
78  Wiggins 1987: 68–69.
79  Wiggins 1976: 151, 1987: 59.
80  Wiggins 1976: 152, 1980: 172–173.
81  Wiggins 1980: 172–173.
82  Wiggins 1980: 175.

83  This matter will be revisited in Chapter 9.
84  Wiggins 1980: 175.
85  Wiggins 1980: 171.
86  Wiggins 1980: 171 (reiterated in Wiggins 2001: 198–199).
87  Wiggins 1980: 171–172.
88  See Wiggins's autobiographical note to this effect in 1996: 229 (also 2001: 198).
89  Davidson 1973.
90  In this, he follows Quine's work in *Word and Object*, Quine 1960.
91  Davidson 1973: 136–139 (see Malpas 2013 for a good overview).
92  See Joseph 2011 for an overview.
93  Wiggins 2001: 198. See Wiggins 1980: 222 and 2001: 198 for the explicit connection he draws between his own analysis and Davidson's.
94  Wiggins 1987: 69, 1996: 245.
95  Wiggins 2001: 198. For an indication of the shift in focus of Wiggins's analysis of 'person', contrast this rendering with the claim that 'a person is, *par excellence* (and as a presupposition of all the traditional questions in the philosophy of mind), the bearer of *both* M-predicates *and* P-predicates' (1987: 63).
96  Wiggins 2012: 16.
97  NB Not a *definition*. Wiggins 1987: 68–69.
98  Wiggins 1987: 71.
99  Wiggins 1980: 222 (a point he gets, perhaps, from Grandy's emphasis on 'humanity' – see Wiggins 1980: 222).
100  Wiggins 1980:

> [W]e have to start to envisage radical interpretation proceeding through a succession of approximations which, at some points, will involve holding belief relatively constant in the light of what the world represents to the subject's experience *taken together* with the concerns of the subject.
>
> (222)

101  Wiggins 1980: 222 – see also Wiggins 1987: 70–71.
102  Wiggins 1996: 245.
103  Wiggins 1996: 248, 2000: 1.
104  Wiggins 1987: 71.
105  Wiggins 1987: 72. There is a stronger version of this argument from interpretation in Wiggins's paper 'Truth, Invention, and the Meaning of Life' (1988). There he suggests that we can only get 'on net' with others if we have the same neurophysiological make-up – so non-human persons are excluded (though artificially synthesized humans are not). He writes:

> Surely neither the consensual method nor the argued discussion of such forms would be possible in the absence of the shared neurophysiology that makes possible such community of concepts and such agreement as exists in evaluative and deliberative judgments. Nor would there be such faint prospects as there are of attaining reflective equilibrium or finding a shared mode of criticism.
>
> (Wiggins 1988: 134, fn. 53)

See also Nicholas Rescher for a similar view (Rescher 1982: 37).
106  Wittgenstein 1953: §206.
107  Wiggins 2001: 199.
108  Wiggins 2001: 199.
109  It is worthwhile noting that this conceptual consilience also seems to immunize Wiggins against the sort of 'thinking part puzzle' described by Olson (e.g. 2007: 215). Understood as concordant with *human being* (a natural substance), our substance sortal predicate 'person' will not apply to undetached heads, brains, etc. (cf. Olson 2007: 218). There is thus no reason to pursue the kind of eliminativist picture

encouraged by Olson and by Merricks (2001). These concerns are examined in greater depth in Chapters 7 and 8.

110 Wiggins 2001: 197.

111 Wiggins 2001: 202, 2005a: 475.

112 As described by Olson (1997: 10).

113 There will, however, be some questions about the extent of the interference of technology (discussed in Chapter 9).

114 Note that a question arises here about the *boundaries* of the human being. When does the principle of activity become realized? Wiggins writes

> Normally the zygote becomes the embryo, but sometimes it divides and becomes twin embryos.... I am committed to react to this fact with a general ruling to the effect that the human being dates from a time after the zygote finally splits or settles down to develop in a unitary fashion.
>
> (2001: 239)

Such issues are returned to in Chapter 6.

## Bibliography

Bakhurst, D. (2005) 'Wiggins on Persons and Human Nature', *Philosophy and Phenomenological Research* 71(2): 462–469.

Butler, J. (1736) *First Dissertation* to the *Analogy of Religion Natural and Revealed to the Constitution of Nature*.

Davidson, D. (1973) 'Radical Interpretation', *Dialectica* 27: 314–328.

Deutscher, M. and Martin, C.B. (1966) 'Remembering', *Philosophical Review* 75(April): 161–196.

Ishiguro, H. (1980) 'The Primitiveness of the Concept of a Person', in Z. Van Straaten (ed.) *Philosophical Subjects: Essays Presented to P.F. Strawson* (Clarendon Press: Oxford).

Joseph, M.A. (2011) 'Davidson's Philosophy of Language', *Internet Encyclopedia of Philosophy* available online at www.iep.utm.edu/dav-lang/.

Locke, J. (1690/1975) *An Essay Concerning Human Understanding*, P.H. Nidditch (ed.) (Oxford: Oxford University Press).

Malpas, J. (2013) 'Donald Davidson', in E.N. Zalta (ed.) *The Stanford Encyclopedia of Philosophy* (*forthcoming*), available online at http://plato.stanford.edu/archives/sum2013/entries/davidson/

McDowell, J. (1994) *Mind and World* (Cambridge, MA: Harvard University Press).

Merricks, T. (2001) *Objects and Persons* (Oxford: Clarendon Press).

Noonan, H. (1998) 'Animalism versus Lockeanism: A Current Controversy', *The Philosophical Quarterly* 48: 302–318.

Olson, E. (1997) *The Human Animal: Personal Identity Without Psychology* (Oxford: Oxford University Press).

Olson, E. (2007) *What Are We? A Study in Personal Ontology* (Oxford: Oxford University Press).

Parfit, D. (1984) *Reasons and Persons* (Oxford: Oxford University Press).

Quine, W.V.O. (1960) *Word and Object* (Cambridge, MA: MIT Press).

Rescher, N. (1982) *Empirical Inquiry* (London: The Athlone Press).

Shoemaker, S. (1970) 'Persons and Their Pasts', *American Philosophical Quarterly* 7: 269–285.

Shoemaker, S. (2004a) 'Brown-Brownson Revisited', *The Monist* 87(4): 573–593.

Shoemaker, S. (2004b) 'Reply to Wiggins', *The Monist* 87(4): 610–613.

Snowdon, P. (1996) 'Persons and Personal Identity', in S. Lovibond and S. Williams (eds) *Essays for David Wiggins: Identity, Truth and Value* (Blackwell Publishing).

Snowdon, P. (1998/2009) 'Strawson, Peter Frederick', in E. Craig (ed.) *Routledge Encyclopedia of Philosophy* (London: Routledge). Available online at www.rep.routledge.com/article/DD066SECT5.

Snowdon, P. (2009) 'On the Sortal Dependency of Individuation Thesis', in H. Dyke (ed.) *From Truth to Reality: New Essays in Logic and Metaphysics* (London: Routledge).

Strawson, P.F. (1959) *Individuals: An Essay in Descriptive Metaphysics* (Methuen).

Unger, P. (1990) *Identity, Consciousness and Value* (New York: Oxford University Press).

Wiggins, D. (1976) 'Locke, Butler and the Stream of Consciousness: And Men as a Natural Kind', *Philosophy* 51: 131–158.

Wiggins, D. (1980) *Sameness and Substance* (Cambridge, MA: Harvard University Press).

Wiggins, D. (1987) 'The Person as Object of Science, as Subject of Experience, and as Locus of Value', in A. Peacocke and G. Gillett (eds) *Persons and Personality* (Oxford: Blackwell).

Wiggins, D. (1988) 'Truth, Invention, and the Meaning of Life', in G. Sayre-McCord (ed.) *Essays on Moral Realism* (Ithaca: Cornell University Press), 127–165.

Wiggins, D. (1996) 'Replies', in S. Lovibond and S. Williams (eds) *Essays for David Wiggins: Identity, Truth and Value* (Oxford: Blackwell Publishing).

Wiggins, D. (2001) *Sameness and Substance Renewed* (Cambridge: Cambridge University Press).

Wiggins, D. (2004a) 'Reply to Shoemaker', *The Monist* 87(4): 594–609.

Wiggins, D. (2004b) 'Reply to Shoemaker's Reply', *The Monist* 87(4): 614–615.

Wiggins, D. (2005) 'Reply to Bakhurst', *Philosophy and Phenomenological Research* 17(2): 442–448.

Wiggins, D. (2012) 'Identity, Individuation and Substance', *European Journal of Philosophy* 20(1): 1–25.

Wittgenstein, L. (1953) *Philosophical Investigations* (Oxford: Basil Blackwell).

# 5    A genealogy of the *person* concept

The variety of functions that 'the' concept of a person plays – the variety of conceptions of personhood we have – cannot be plausibly combined in a single concept. At most, one might settle for a heterogenous class, defined by a disjunction of heterogenous conditions. Even if some rough construction of a denominator common to all of these notions and functions were proposed, that conception would be so general that it could not fulfil – nor could it generate – the various functions performed by the various regional and substantively rich conceptions.[1]

Descriptivists like Strawson and Wiggins turn to our conceptual framework to extrapolate the structure of reality. We have seen that this method is not without its problems. Among them is the concern raised by E.A. Burtt and Tsu-Lin Mei that in examining their 'everyday thoughts' the descriptivist runs the risk of importing cultural bias into their metaphysics. In this chapter, I contend that Wiggins's analysis of our use of the term 'person' does precisely this. He holds that his *semantic analysis* will grant us insight into the pre-theoretical concept that supposedly sustains it. The claim set out below is that our use of the term is guided by diverse and sometimes conflicting rationales and that it does not, therefore, provide a stable basis for his descriptivist claims. The fragmentary nature of our notion of a 'person' becomes clear when it is subjected to a genealogical analysis.[2]

In some ways, under several heads, this chapter falls on a different register from the rest of the work in this book. Despite the disclaimers about historical scholarship I entered at the start of Chapter 1, it is a historical critique. It is important to emphasize, therefore, that my failings as a historian of ideas are compensated (I hope) by the virtuosity of others – not least Amélie Oksenberg Rorty (from whom I take the quotation above) and Marcel Mauss. The account given below is in need of refinement,[3] by someone more capable than I am; nonetheless, I think the critique still stands and offers a real objection to Wiggins's approach to the question of personhood.

## 1  The notion of a person

'Genealogy', as a philosophical method, came to prominence in the 1970s in the work of Michel Foucault[4] (himself drawing largely on the work of Nietzsche[5]).

It has since been picked up in the Anglophone tradition by Bernard Williams,[6] Miranda Fricker,[7] Quentin Skinner,[8] and Raymond Geuss.[9] Crudely put, genealogy is a form of historical critique, designed to problematize norms by presenting them, not as natural and inevitable, but as products of diverse historical factors.[10]

Genealogical analysis is used in a variety of ways, but for present purposes the 'modified' genealogy that Geuss deploys is perhaps most suitable. In *Public Goods, Private Goods*, he presents genealogy as taking as its object of study some 'deeply entrenched contemporary item or phenomenon' – in this case a conceptual distinction between 'public' and 'private'.[11] For Nietzsche it was Christian morality, for Foucault it was the prison system and sexuality. In advance of the application of genealogy each of these items presents itself as coherent. While there may be different elements, each 'item' appears to be unified by some single rationale.[12] Furthermore, this rationale is often construed as stemming from a single point of origin. Thus sexuality is supposed to have its natural origin in biological nature, and Christianity, despite its schisms and heresies, 'can still be seen as arising from a unitary aboriginal *Sinnstiftung* by Jesus in Palestine two thousand years ago'.[13]

Wiggins recognizes that we use the term 'person' in different ways, but throughout his work he tries to show how, ultimately, these different conceptions are bound together by a unitary rationale. This aim is made explicit in the 1987 paper, 'The Person as Object of Science',[14] where he draws the connections between the three seemingly disparate aspects of our everyday thoughts about 'persons':

1   the idea of the person as object of biological, anatomical and neurophysiological inquiry;
2   the idea of the person as subject of consciousness;
3   the idea of the person as locus of all sorts of moral attributes and the source or conceptual origin of all value.[15]

There is something admirably optimistic about the kind of story Wiggins tells, presenting 'person' as capturing a coherent idea in spite of the variety of ways that we apply the term. And there is something rather sad about trying to disassemble this jigsaw and show that some pieces will never fit. But such is the aim of the genealogical project (certainly as Geuss conceives it). A genealogical analysis shows how such ideas, far from being unitary, are in fact syncretic phenomena, with many distinct roots, no single beginning (hence the 'genealogical' metaphor) and no unifying rationale. The elements of Christianity, born from Stoic thought, Roman law, Hebrew scriptures and the politics of resistance to Roman power (etc.) are distinct and possess completely different rationales.[16] The same, I shall argue, is true of the notion of a 'person' – and this should be no surprise, since the process by which that concept reached its current state will be just as complex. (Indeed it encompasses Christian doctrine, relating as it does *inter alia* to the soul, and the Holy Trinity.[17]) As Geuss puts it:

Any significant human phenomenon that has succeeded in maintaining itself throughout a long history into the present ... can be expected to be a highly stratified composite whose parts derive originally from different periods. The original rationale of each of these parts will have been oriented to a completely different (past) context of action.[18]

The notion of a 'person' is just such a phenomenon.

Wiggins claims that *person* and *human being* are non-accidentally concordant. His analysis of the semantics of 'person', and its associations with the notion of a subject of interpretation, is meant to secure the connection between the two concepts – but the genealogical approach problematizes any such treatment by suggesting that our use of the term is informed by numerous, culturally diverse, and sometimes conflicting influences, unbound by a unifying rationale.

Three preliminary points should be made before we start. First, the aim of our approach is not to discredit a practice, institution or item simply by focusing on a shameful origin. It is not party to the genetic fallacy. Indeed it presupposes *no* originary point.[19] Second, like Geuss's, the method of genealogy is here 'modified' in the sense that it is not directed at an institution or practice, but an idea.[20] The aim of my analysis will be to test the appearance of unity in our everyday notion of a person.

Third, a genealogical study may seem to be an overly ambitious project and one ill-suited to the space constraints of the present study. But my aim is modest. The story entered here is intended only as a rough sketch of *some* of the various disparate elements of the notion of a person. And insofar as it is ambitious then the ambition exists outside this book. There are two texts that directly inform the analysis: Amélie Oksenberg Rorty's excellent 'Persons and *Personae*'[21] and Marcel Mauss's 'A Category of the Human Mind: The Notion of Person; the Notion of Self'.[22] Both attempt to show, in different ways, how 'the various functions performed by our contemporary concept of person do not hang together'.[23] They suggest the general value of such a genealogy outside the current discussion. Mauss's anthropological study will be particularly important.[24]

Mauss's last essay, published in 1938, is a survey of various historical and ethnographical studies in which he describes the long and complex history of the seemingly natural notion of a person. He identifies its different renderings and charts how they came into contact and were subsequently conjoined, moving from societal role to legal fact, to moral fact and to metaphysical entity:

From a simple masquerade to the mask, from a 'role' to a 'person' to a name, to an individual, from the latter to a being possessing metaphysical and moral value, from a moral consciousness to a sacred being, from the latter to a fundamental form of thought and action.[25]

This quotation gives a sense of the vastness of the project. It also indicates one of the problems with Mauss's approach. Emerging from the French sociological school dominated by Durkheim, it appears to construe the distinct notions as related in some form of linear narrative. The different points appear as stages which lead on from one to the next. This, it should be noted, is troublesome for the genealogist, who sees her approach to be characterized by multiple branching (or, in Deleuzian idiom, as 'rhizomatic').[26] It is also potentially politically suspect, since this kind of evolutionary picture positions some cultures as 'less evolved' than others.[27]

Whether Mauss intended his narrative to be linear is unclear.[28] Either way, my aim here is to show how it can be read as something like a genealogy. I build on the different strands that he identifies and the interrelations that he sees between them in order to raise questions for Wiggins. What follows may be seen as a creative *reinterpretation* of Mauss, drawing out by reference to his cross-cultural study the different components of the notion captured by the word 'person'.

This being said, there is still no better introduction to the essay, or these thoughts, than Mauss's own – and the relevance to Wiggins's work is nowhere better signalled than at the start of 'A Category of the Human Mind':

> [This essay] deals with nothing less than how to explain ... the way in which one of the categories of the human mind – one of those ideas we believe to be innate – originated and slowly developed over many centuries and through numerous vicissitudes, so that even today it is still imprecise, delicate and fragile, one requiring further elaboration. This is the idea of 'person' (*personne*), the idea of 'self' (*moi*). Each one of us finds it natural, clearly determined in the depths of his consciousness, completely furnished with the fundaments of morality which flows from it. For this simplistic view of its history and present value we must substitute a more precise view.[29]

## 2  'Person' as a mask

Mauss's investigation begins with the 'mask' culture of the Native American Pueblo of Zuñi, and the pre-Roman Etruscans; he finds in their practices the notion of the 'person' as something like a *mask* or *role*, correlating to the way we still talk of the *dramatis persona*.[30] And while the practices of the Pueblo of Zuñi are far removed from the historical process out of which the 'personal identity' debate emerges, this use of person is – as many have noted – apparent in the etymological roots of the English word 'person', as a mask, through which (*per*) resounds the voice (*sonare*). (Wiggins himself recognizes this.)[31]

In these 'mask cultures', the communities or 'totemic groups' are described as possessing only a fixed number of names.[32] In naming rituals, these 'characters' or 'roles' or 'persons', are conferred on members of the group. Following this baptismal rite, the bearers of the name are regarded, at any time, as the *reincarnation* of the original bearer.[33] Indeed, for Mauss, this account of 'person' as mask is central to the logic of reincarnation, where a single person is seen to span different biological lives:

The individual is born with his name and his social functions.... The number of individuals, names, souls and roles is limited in the clan, and the line of the clan is merely a collection (*ensemble*) of rebirths and deaths of individuals who are always the same.[34]

On first reading, these kinds of ethnographic details might seem to register on a completely different level from the one that Wiggins is working on. But attention to *S&S* and his earlier paper, 'Locke, Butler and the Stream of Consciousness', reveals that he explicitly engages with exactly this kind of hypostatization of social roles, where he analyses our understanding of what it is to be a 'person'.[35] His point of reference is Clifford Geertz's *Person, Time and Conduct in Bali* rather than Mauss, but the issues discussed are strikingly similar, focusing as they do on 'the strange fusion of role and human being that is involved in [the Balinese] system of naming'.[36]

Wiggins's interest in Balinese naming practices surface in his discussion of fission. On the neo-Lockean model fission becomes a conceptual possibility because any theory that puts an exclusive focus on consciousness is placed to rule against, for example, hemispherectomies and subsequent 'deltas' in the stream of consciousness.[37] One aim, in 'Locke, Butler and the Stream of Consciousness' and *S&S*, is to demonstrate how this sort of splitting would destabilize our *everyday usage* of the term. It involves thinking of persons as things that can *transcend bodies* and *individual lives*.[38] If this is right, then for the descriptivist the neo-Lockean's memory and consciousness cannot constitute personal identity (since it clashes so forcefully with our everyday pre-theoretical practices). He writes:

> The conceptual possibility of a delta in the stream of consciousness jogs our whole focus on the concept of personhood.[39]

Geertz's study of Balinese society becomes important, therefore, because it describes an 'extant conception',[40] where *person* functions less like a substance sortal and more like what Wiggins calls a 'concrete universal'.[41]

In his early work, Wiggins's notion of 'concrete universals' is explicated by a figure drawn from Plato's *Parmenides*.[42] The idea is of a type of being with potentially spatially and temporally dislocated parts – the collection of all Cox's Orange Pippin trees, for example,[43] or, in this context, the collection of those who wear the same mask.[44] (Whatever they are, *concrete universals* have a different metaphysical character from *substances*.) For the Balinese, a single person may be comprised of distinct individuals. Nor is this hypostatization of social roles restricted to the Balinese, the Zuñi and the pre-Roman Etruscans. Wiggins finds a version of this concrete universal conception of person in our own culture (or, as he puts it, 'a neighbouring compartment'). He quotes from Sartre's commentary on Flaubert:

> *Un homme n'est jamais un individu; il vaudrait mieux l'appeler un universal singulier; totalisé et, par la même, universalisé par son époque, il la retotalise en se reproduisant en elle comme singularité.*[45]

[A man is never an individual; it would be better to call him a singular universal; total, and also, universalized by his epoch, he retotalizes it by reproducing it as a singularity.]

Examples abound in Western literature.[46] The thought is that in these contexts 'person' is understood as a role that is played by different individuals. It appears that it is not only an 'extant conception', but one can seem to persist in some of our own thoughts about *what persons are*. Yet, as Wiggins notes, this clashes with the simultaneous and persuasive thought that a person is an entity with an individual biography:

> [T]here is no question [on this view] of building up a coherent historical record of the individual passions, thoughts and actions of an individual person. There is little or no provision for the individual or, as it were, the perspectival aspect of human experience. The whole ordering of the events of human history is interpreted so far as possible in terms of the recurrence of generic types of doing or suffering. And where there scarcely is such a thing as history, the idea of biography loses all purchase.[47]

Wiggins sees the concrete universal conception as clashing too strongly with our everyday attitudes to person and the practices that relate to them. He wants to shear it from the central notion. Indeed he has come to see involvement in this idea as a *reductio ad absurdum* of any theory of personhood that is party to it. But I think much more needs to be said. The clash between the 'mask' conception of person and our everyday practices may be taken as grounds for rejecting that conception of person; but could it not equally be taken as evidence of the contradictory make-up of the notion itself? It is true that the 'concrete universal' conception sits awkwardly alongside our use of 'person' to pick out a locus of moral attributes, but maybe that is because our notion is syncretic, comprising elements deriving from diverse rationales. It is not immediately clear what licenses Wiggins's disavowal of the 'concrete universal' conception. Maybe it is significant that this discussion of Geertz and the Balinese disappears after *S&S* (and is never referred to in the secondary literature).

## 3 'Person' as a legal fact

I come next to another potentially controversial element. Mauss sees the 'mask' rendering of person to be one of the earliest stages in its evolution. He identifies evidence of this kind of 'mask' civilization in early rituals in Rome,[48] but then he describes a conceptual move away from the person as 'a mask, a tragic mask, a ritual mask and the ancestral mask', towards the *privileges* of those individuals with a *right* to a mask. Personhood becomes 'a fundamental fact of law'.[49] Rights are accorded to the bearers of masks (and then to those holding names which stand in the place of masks).[50] Thus compare the status of Roman citizens or 'legal persons' ('all freemen of Rome were Roman citizens, all had a civil *persona*'[51]) with their slaves ('*Servus non habet personam*'[52]), beings who lacked 'personality', owning neither body, nor ancestors, nor name, nor belongings. They are individuals but not persons. Here, the person is presented as a legal entity and 'personhood' is a legal status. The individuals who are granted it are subject to the laws of persons rather than of property.[53]

This legal sense of person still survives. It marks a connection between legal thinking and the 'mask' conception that in the modern juridical usage different individuals may still count as the same person. For, in the current British legal system *corporate bodies* register as 'persons':

> **IV.6.** *Law. a.* 'A human being (*natural person*) or body corporate or corpo-ration (*artificial person*), having rights and duties recognised by the law ... *TUCKER Lt. Nat. (1834) II. 188 A crowd is no distinct existence ... but if the same people be erected into a corporation, there is a new existence superadded; and they become a person in law capable to sue and be sued* [etc.]'[54]

Here, then, we find a point of tension between two ideas about what persons are. This legal conception sees our usage of the term 'person' to refer to some (legally specified) *stipulative definition*, which is palpably at odds with Wiggins's claim that 'person' functions like (although it might not *be*) a natural kind word.[55] 'Person', here, is a construct of legal practice – a being with certain rights and duties – and we learn about persons in this sense, not by looking to the world but by looking to how they're specified in law (in the same way as we grasp what a chairman is, for example).

Here, the marks of personhood are seen to be socially determined. The concept of the person is 'constructed' (hence we call those who hold this posi-tion 'constructionists'). What counts as a person is a matter of convention or tacit agreement. It has a nominal essence that is up for discussion.[56] That we use 'person' in this way is noted by Snowdon, in his contribution to *Identity, Truth and Value*:

> [I]t is not clear that there is any deep difference between those concepts which we all regard as capable of nominal elucidation and the notion of a person.[57]

Snowdon rejects the natural kind reading in favour of the stipulative one,[58] but my suggestion is that both are valid. Our present notion is a stratified composite with conflicting elements.

Indeed, Wiggins does not always deny that 'person' can be read as specifying a nominal essence.[59] In 'The Person as Object of Science' he writes:

> [T]he concept *person* had best not be like the concepts of various kinds of executant or like concepts that have to change in response to technological progress such as *surgeon*, or *infantryman* or *footplateman*, or like legal

concepts that we may decide at any moment to modify, such as *tenant, citizen, metic, minor,* or like legal concepts that we simply invent, such as *patrial.*[60]

He does not deny that we *do* sometimes use 'person' in this way. Yet he warns us against such usage because he thinks that to treat 'person' in this way is to invite a form of moral degeneration. There are social and political implications of treating the concept of *person* otherwise than as a peculiar restriction of a natural kind concept. This view is made explicit towards the very end of *S&S*:

> Not only does [a constructionist conception of persons] reduce the theoretical subject matter of morals and politics and limit the range and variety of counterfactual speculation that this subject matter can be expected to sustain; and not only does it reduce drastically (for the same reasons) the scope for real criticism of the actual works of social engineering that ought to have [been] held in check by a healthy respect for the partly imponderable real essence of actual persons: it will also license a state of affairs in which there was absolutely nothing except fear of confusion to obstruct proposals for modifying or reinventing even the *accepted* specification of what a person was – just as we constantly and effortlessly modify and refashion through time certain institution – and artifact-concepts.[61]

This is one section of what Wiggins later called his 'anti-constructionist tirade of 1980'[62] – a furious attack that has largely been cut from subsequent texts (and appears only in condensed form in the concluding pages of *S&SR*).[63] The reason for this cut is not made explicit, but as David Bakhurst notes, many will find it judicious.[64] Some will think the constructionist a straw-man.[65] Others, like Bakhurst, will find something awkward in this moral attack on a metaphysical position.[66] To these lines of criticism the following can be added: irrespective of its damaging outcomes, the descriptivist's concern should be with how we *use* 'person', and in some context we certainly use 'person' as though it possessed some nominal essence; moral objections to this usage will not show that it is incorrect, only that it is problematic.

## 4 'Person' as moral fact and metaphysical entity

Let us finish Mauss's story. The sense of person that grounds Wiggins's account only starts to appear towards the end of the narrative. Mauss describes a third strand in the idea of a person, the result of the interweaving of the juridical notion and the legal-cum-moral '*prosopon*' that is found in the Stoic tradition. What results, from the cross-pollination of Greek and Roman thought, is a notion of person as both the legal 'mask' and also the *true face* of the person/citizen.[67] Here, a person is both a role or position with particular rights and duties, and also the individual who is responsible for their own actions, who is faced with choices, with respect to those legal duties:

The word πρόσωπον did indeed have the same meaning as *persona*, a mask. But it can then also signify the 'personage' (*personnage*) that each individual is and desires to be, his character (the two words are often linked), his true face ... πρόσωπον is no longer only a *persona*, and – a matter of capital importance – to its juridical meaning is moreover added a moral one, a sense of being conscious, independent, autonomous, free and responsible.[68]

As Mauss points out, we find the superimposition of the juridical and moral aspects in Marcus Aurelius' command to 'carve out your mask'[69] – and this connection, between the legal and the conscious *moral* person transfuses more recent accounts. It is a point bordering on the banal, but it is no surprise that Locke's account of personhood is constructed by reference to the legal conception.

Even here, however, where we are moving into more familiar territory, 'person' will not carry the full weight that Wiggins assigns to it. In particular, the persona/*prosopon* conception as it stands in the Greco-Roman tradition does not refer to *each and every* human being. It still functions as a status, one which can be achieved or withdrawn. As Martin Hollis points out, in the Classical era there was 'no single generic word for each and every human being'[70] (to say nothing of those cultures perhaps further from Wiggins's own, such as the Balinese and the New Guinean).[71,72]

It is only after the advent of Christianity that Mauss sees 'person' as picking out each and every human being for the entirety of their existence. It is here that we have the transition from 'the notion of *persona*, of "a man clad in a condition", to the notion of man quite simply the notion of the human "person" (*personne*).'[73] It is this fourth strand of the notion of a person that grounds Wiggins's treatment. There is no need for present purposes to scrutinize it fully. One need only say that, following the intermingling of the Christian doctrine of the Trinity and the evolution of the concept of the soul, Mauss claims that 'person' comes to apply to all human beings. Just as all human beings come to have a soul, 'which Christianity had given them'.[74] Personhood becomes something that humans, citizens or slaves cannot be stripped of. This is the understanding of *person* that supports Wiggins's study.

## 5  Critique of the semantic analysis

As we use it in everyday speech, 'person' functions in multifarious ways. The analysis above is taken to lend substance to the claim that the notion of a 'person' is a cultural accretion, a 'syncretic phenomenon'. It is not, as Mauss presciently saw, a 'primordial innate idea, clearly engraved since Adam in the innermost depths of our being....'[75] Our use of the term 'person' is guided by *conflicting* rationales. We use it as though it picks out particulars but also, sometimes, as though it picks out concrete universals; we use it as though it were a natural kind word, but also stipulatively. The thought here is that an analysis of our use of so plastic a term cannot provide a stable basis for Wiggins's descriptivist claims. How we use it will grant us no insight into a coherent pre-theoretical concept.[76]

Furthermore, focusing on 'our' use of the term may be seen to introduce cultural biases into Wiggins's descriptivist picture. Wiggins is interested in some strands of the notion of a 'person' but not others. The concrete universal conception is marginalized – but what then of cultures, such as the Balinese, which put particular emphasis on this understanding of personhood? Do they fail to capture a fundamental element of our conceptual framework? Wiggins can plausibly avoid such morally and politically disturbing conclusions[77] by revising his reading of Geertz, and focusing on the obvious capacity that he shares with the Balinese for getting 'on net'. It is important all the same to recognize the worry which touches all metaphysical inquiries, but achieves particular prominence with the descriptivist project, that the cultural bearing of the philosopher will have direct effects on the metaphysical claims he or she is led to make.

Do these thoughts disturb the conceptual consilience of *person* and *human being*? It should be remembered that Wiggins does not think there is a 'transcendental argument' to conjoin the two concepts.[78] He simply aims to provide arguments that strongly suggest a connection. And where the semantic analysis fails, he may turn to the other two connecting cords: the Strawsonian argument and the argument from interpretation. Perhaps they more neatly tie *person* and *human being* together?

Perhaps – but if so, the connection is not quite of the kind that Wiggins claims.

The Strawsonian argument and the argument from interpretation are not based on how we use the term 'person'; rather, they are used to flesh out the structure that undergirds our everyday thoughts. If we examine our modes of thought ascription we will find that thinking things must be material things. If we look to how we interact with one another we see that we must understand others as *subjects of interpretation*. Wiggins takes these thoughts to elucidate an element of our conceptual framework: the *person* concept.

These are persuasive arguments. Each gestures towards a conceptual concordance between the basic, pre-theoretical thought and the thought of a *human being* – 'one of *us*', the kind of biological being with our distinctive capacity for culture and so on. It is on these grounds that Wiggins holds that *person* and *human being* assign the same *principle of activity*. When we are interested in the persistence conditions of the one, we can examine the persistence conditions of the other. Claiming this, he is implicitly committed to the thought that both *person* (which we might reasonably take to cover the things whose persistence the personal identity debate is focused on tracking) and *human being* are *substance sortals*. The question to be investigated in the remainder of this chapter is: what justifies this substance sortal reading of *person*?

'Person' looks like a fundamental or basic sort of classification.[79]

This conclusion emerges, in 'The Person as Object of Science', from Wiggins's semantic analysis. When we use the term 'person', we use it as a substance sortal predicate, which applies to an individual for every moment of their existence.

Yet our everyday usage is not, as noted, a stable foundation for Wiggins's position. So, can the substance sortal reading be substantiated by the other two arguments?

On a generous interpretation, the Strawsonian argument supports the claim that certain psychological states can only be enjoyed by beings with a particular physiology – the human one. But this, by itself, is not enough for Wiggins's purposes. It might be basic to our conceptual framework to understand psychological beings as material beings; but it remains open whether or not psychological beings must be psychological beings for every moment of their existence.[80]

What about the analysis founded in Davidson's ideas about interpretation? Is the notion of a *subject of interpretation* necessarily the notion of a fundamental sort of thing? If an individual is a subject of interpretation is it necessarily so at every moment of its existence? Consider again what is meant by 'interpretation':

> [T]he only way for us to make sense of being a person is to think of persons … as subjects of fine-grained interpretation by us, as subjects for whom we are subjects of fine-grained interpretation, and as creatures with whom we can have relations of co-operation and reciprocity.[81]

Understood thus, can we conceive of personhood as something *acquired*? Here, let us turn to our everyday dealings. Think of gestating or newly born babies. Think of the way we interact with them. We do not treat them as beings we can reason with, whom we can engage in joint projects. We do not see them as beings who – but for some physiological impairment – interpret, reason, empathize or imagine themselves in our shoes (unlike, for example, adults suffering severe mental delay). There are conceptual capacities they have yet to acquire.[82] There is potential there, but we do not enter into co-operative and reciprocal relations with them.

The thought that personhood is acquired is nicely captured by David Bakhurst in his commentary 'Wiggins on Persons and Human Nature'.[83] Bakhurst emphasizes the process of *enculturation* that eventually allows a child to 'tune in' or 'get on net', with those around it:

> Human beings owe their distinctive psychological powers to 'cumulative cultural evolution': each individual inherits the legacy of the collective achievements of past generations. This is not just a matter of the transmission of knowledge and skills across generations. Rather, children's minds and characters are formed through their assimilation of culture. At its strongest, this perspective endorses an idea that might be thought anathema to Wiggins, though it emerges in the work of thinkers he admires, such as John McDowell and Lev Vygotsky: the idea that personhood is acquired. On such a view, human beings are born 'mere animals' and become persons as they attain a 'second nature', in the form of conceptual capacities and moral sensibilities which make them, as McDowell puts it, 'at home in the space of reasons'. These capacities and sensibilities are acquired and refined

through *Bildung* – initiation into the traditions of thought and action embodied in language and culture. Personhood is thus both a result of the normal maturation of human beings and an artefact of culture.[84,85]

Bakhurst's proposal can be read as suggesting that the *person* concept – underwritten by the Davidsonian notion of a *subject of interpretation* – need not be a notion of a fundamental kind of thing. And this seems to resonate with our interpretative procedures; it makes sense, for example, of many of the ways that we think of and interact with gestating or newly-born babies. At the very least it puts the onus on Wiggins to explain why, if one is a subject of interpretation, one must have always have been so.

Wiggins's reply to Bakhurst is as follows:

> Reading his proposal quite literally, I am troubled to think that one might find oneself saying that A was more of a person aged 32 than he was at 20, or that B was more of a person than C was and meaning it quite literally. Once we started saying that sort of thing and taking it as seriously confirming Bakhurst's proposal, we should need to take more seriously than we now do the fact that we sometimes say 'D is a real Mensch' whereas (we say) 'E is scarcely human' – ways of talking we do not think of as committing us to think of human beinghood as something acquired or committing us to exempt E from all reproof for his callousness or brutality. Is it not better to conceive of a human person as a creature with a *natural capacity*, which may or may not be realized, for reason, morality, Bildung ... and better to say that these achievements *fulfil the potentialities* of human beinghood/personhood? Adapting a dictum of Woodger's I have quoted before would it not be better to say that the child is the *primordium* of the moral/rational being but not of the future person, because the child already is that person?[86]

Yet Bakhurst's proposal is in conflict with none of this. Wiggins is articulating two objections; neither is an insurmountable obstacle.

His first objection sounds on a moral register and follows from his views, alluded to above, about what it means to be a 'good person'. In his 1976 essay, 'Truth, Invention, and the Meaning of Life',[87] he suggests that to think of personhood as a status to be *achieved* 'make[s] a mess of what we mean by "a good person"'.[88] Where personhood is seen as a socially-determined status, 'person' is used as a functional term (i.e. as referring to beings with a determinate, specifiable function/purpose) and this is at odds with the emphasis in our everyday moral language, which revolves around notions of autonomy and the exercise of our human capacities for self-direction.[89] This is one way of understanding the claim above that it is 'better to conceive of a human person as a creature with a natural capacity, which may or may not be realized, for reason, morality, Bildung ...'. The pre-theoretical concept is deeply embedded in our moral practices – and our moral practices belie any *functional* reading of it (as having a *principle of functioning*).

Wiggins's second and related point seems to be that persons are better understood as creatures with certain natural capacities because it is only by seeing them thus – and not as beings defined by some nominal essence – that we can accommodate the *aposiopesis* that is a necessary correlate of seeing individuals as *subjects of interpretation* (note how the ellipsis that signals the essential incompleteness of the list is included in the quotation).[90] We cannot stipulate what capacities a being must have to be 'one of us'. We must look to our experience of the world to discover this. When thinking properly 'about what interpretation and reciprocity involve on the levels of reason and response',[91] the only stereotype we can have of a person is a human being.[92]

Yet Bakhurst can wholeheartedly agree that the constructionist, stipulative account of personhood is problematic. His suggestion, refined here, is only that our stereotype of a person is a human being who has achieved some level of psychological-cum-cultural integration (not, for example, an unborn, or newly born, baby). When we think about persons we naturally think of beings whom we can interpret and who can interpret us; our stereotype is a human being who has passed through certain developmental stages. (Wiggins himself exhibits this bias in the above quotation, where he suggests that Bakhurst's proposal licenses the thought at a 32-year-old could be 'more of a person' than a 20-year-old. It might suggest that a 32-year-old is more of a person than a one-day-old, or an unborn baby in the third trimester of gestation – but this is less intuitively surprising.) Seeing personhood as a stage in the life of a human being does not commit one to thinking that the marks of personhood are stipulated by us. We do not think of *butterfly* or *fawn* as being nomologically shallow concepts just because they refer to stages in an animal's life.

Perhaps Wiggins will say – following Aristotle – that human beings are essentially 'rational animals'. On these grounds he may justifiably object to Bakhurst's claim that 'human beings are born "mere animals" and become persons as they attain a "second nature"' since it conflicts with the Aristotelian thought that rationality pervades our being and is not a skill to be acquired.[93] And, in his early work, Wiggins does appear to interpret 'person' along these lines, as being the form, the *psuche*, of the human being. He writes, in the avowedly Aristotelian *ISTC*:

> [F]or our purposes it will not do very much harm to think of *psuche* as much the same notion as *person*.[94]

Yet this is not how he reads 'person' now. Roughly, a *rational animal* is a creature that can learn, and understand, and think about itself, and find its place in the world. A *person* is such a creature that is also *one of us* – a being who we can interpret and who interprets us. As was seen in Chapter 4, there may be rational animals – dolphins or aliens, for example – who are *not* persons.[95]

Where does this leave us? Nothing that has been said seems to have severed the conceptual connection between *person* and *human being*. Leaving the semantic analysis by the wayside, the Strawsonian argument and the argument from interpretation provide strong grounds for thinking that our pre-theoretical concept – '*person*' – is sustained by our notion of a human being. But I contend that the connection is different in kind from the one that Wiggins finds. In *S&S* and *S&SR*, he holds that both sortals assign the same underlying principle of activity;[96] for this to be the case, both need to be sub-stance sortals. Guided by Bakhurst, however, we are led to the thought that personhood – insofar as it is underwritten by the Davidsonian notion of a subject of interpretation – can be *acquired*. I claimed also that the Strawso-nian argument does not justify the substantial reading. *Person*, while still con-strued as a fundamental element of our conceptual scheme, appears as a restriction of *human being*.[97]

The thought that *person* is a 'fundamental or basic sort of classification' is grounded in the way we use the term 'person'. It is true that we do use it as a substance sortal (think how awkward we find the claim that human beings *become* persons). But we can also use it to describe a *phase*. The 'mask'[98] and 'legal'[99] conceptions of *person* – each residually still available – construe 'person' as a status that an individual can attain. Similarly, on the Lockean account, one may conceive of a man ceasing to be one person and becoming another,

> [This explains] our way of speaking in English when we say such an one is 'not himself', or is 'beside himself'; in which phrases it is insinuated, as if those who now, or at least first used them, thought that self was changed; the selfsame person was no longer in that man.[100]

The genealogy I have set out shows that our uses of 'person' are supported by different, sometimes conflicting rationales. Our use cannot therefore be a stable foundation for Wiggins's analysis. It also indicates a point at which the term 'person' began to function properly as a substance sortal term – after the advent of Christianity. My suggestion is that it is this seam in the rich and variegated notion, and not some pre-theoretical concept, that Wiggins is tapping when he presents his substantial reading of 'person'.

\*   \*   \*

The aim of this chapter has been to suggest a modification to Wiggins's claim about conceptual consilience. Central to his view is a careful and complex eluci-dation of the *person* concept; he shows how it functions in our everyday lan-guage and ties it to the Davidsonian notion of a *subject of interpretation*. What emerges from his analysis is the claim that *person* is conceptually consilient with *human being*. The genealogy, however, encouraged us to test the *semantic ana-lysis* of our use of the term 'person'. Our use, it was suggested, is guided by con-flicting rationale; as a result, Wiggins's reading of 'person' was seen to be inflected by certain socio-cultural biases. Among these was the thought that our notion of a person is necessarily the notion of a *substance*. This thought is not licensed or corroborated by the Strawsonian argument or the argument from interpretation. It might be that our understanding of what it is to be a psychological being involves understanding that being as in some way material; and perhaps our notion of a person *is* developed through examination of creatures on the same wavelength as us. But these claims are compatible with the further thought that *person* is a sortal term that marks a *phase* of the life of a human being. Thus the conceptual relation between *person* and *human being* – if there is one – might well be an asymmetric one. A human being may be understood to *become* a person; their spatio-temporal limits, therefore, are not the same. This undermines a central tenet of Wiggins's Human Being Theory, that those two concepts assign the same *principle of activity*.

## Notes

1 Rorty 1990: 35 (see, for example, Adam Morton for a similar view: Morton 1990).
2 In the following chapter, talk of 'the concept *person*' will be largely eschewed in favour of 'the notion of a "person"', 'notion' being a much less loaded term in Wiggins's texts (and one suggested by Mauss's essay).
3 Giovanni Boniolo (2013) offers some helpful comments on this topic, which a fuller treatment would incorporate (see, particularly, his second footnote).
4 E.g. Foucault 1994.
5 E.g. Nietzsche 1887/1996 (see Foucault 1994).
6 E.g. Williams 2002.
7 E.g. Fricker 2007.
8 E.g. Skinner 2002.
9 E.g. Geuss 1999: 1–29, 2001.
10 See Hill 1998 for an overview.
11 Geuss 2001: viii.
12 Geuss 2001: viii–ix.
13 Geuss 2001: ix–x.
14 Wiggins 1987. A revised version of this paper appears in his latest collection (2016) but rather than contend with the subtleties of these revisions I will focus my attention on the original.
15 Wiggins 1987: 56.
16 Geuss 2001: xi.
17 See, for example, Mauss 1938/1985/1985: 19–20.
18 Geuss 2001: xiii.
19 Geuss 2001: xv.
20 Geuss 2001: xvi.
21 Rorty 1990.
22 Mauss 1938/1985/1985.
23 Rorty 1990: 22.
24 It is gratifying that, following our discussions, Wiggins has incorporated a treatment of Mauss into a paper rewritten in his forthcoming collection (*Twelve Essays*, 2016). However, his reading is sufficiently different from mine to bear mention. He begins by quoting Mauss and then comments on it:

> 'Who knows what progress the Understanding will yet make on this matter [of the notion of person or self].... Who knows even whether this "category", which all of us believe to be well founded, will always be recognized as such? It is formulated only for us, among us. Even its moral strength – the sacred character of the human "person" – is questioned not only throughout the Orient ... but even in the

countries where this principle was discovered. We have great possessions to defend. With us the idea could disappear.' At this point in his essay, Mauss celebrated (as I read him) not so much the historic diversity of conceptions of personhood that he had begun by putting in front of his readers as the gradual emergence from these conceptions of a better conception.

('Sameness, Substance and the Human Person' in Wiggins 2016)

What are we to make of this? The use of 'conceptions' suggests that Wiggins thinks the various renderings are simply different perspectives on the same fundamental concept. Fine – but, if he thinks this (as will become clear from the arguments to follow) he cannot turn to the *usage* of the term 'person' to give us insight into this concept (since there are mutually contradictory uses of it).... He cannot unless he thinks (which perhaps he does) that *his* (our?) use, over those of different cultures, is the 'best' (i.e. that it gives us greater insight into the *person* concept). But, given that this devolves into a peculiarly and problematically Anglo-centric perspective, I would resist this line of interpretation.

25 Mauss 1938/1985: 22. Mauss's list of the different aspect of personhood overlaps and cross-cuts Rorty's functions of personhood. Rorty thinks a person can be read, variously as: (i) a recipient of respect (22), (ii) a legal entity (23), (iii) an autonomous agent (25), (iv) as a 'dramatis persona' (a bearer of roles) (28), (v) as a locus of value (30), (vi) as a biological entity (32) and (vii) the conscious subject of experiences (33). She shows, carefully, methodically, how these different functions do not hang together.

26 Koopman 2012.

27 This is evidenced by the construal of the Pueblo of Zuñi as stuck in the 'Totemic stage', from which our own conception of person evolved.

28 Certainly in this essay, it seems he does not. He talks about simultaneous movements (the movement away from the Totemic mask civilizations and the injection of the 'prosopon' concept from Stoic tradition).

29 Mauss 1938/1985: 1.

30 This is an example of the unfortunate consequences of his evolutionary analysis. The still extant Native American cultures are seen to be in some senses 'stuck' in the 'aboriginal state' (Mauss 1938/1985: 4) out of which our own culture emerged.

31 Wiggins 1996: 282, fn. 27. It is worth noting that Mauss questions this reading:

> In reality the word does not even seem to be from a sound Latin root. It is believed to be of Etruscan origin, like other nouns ending in '-na' (Porsenna, Caecina, etc.). Meillet and Ernout's *Dictionnaire Etymologique* compares it to a word, *farsu*, handed down in garbled form, and M. Benveniste informs me that it may come from a Greek borrowing made by the Etruscans, πρόσωπον ('perso').
>
> (Mauss 1938/1985: 15)

Either way, as he emphasizes, it emerges from the Etruscans' mask civilization. (Mauss 1938/1985: 15) (cf. Boniolo 2013, fn. 2).

32 Allen 1985: 32.

33 Allen 1985: 32.

34 Mauss, as quoted in Allen 1985: 33.

35 Wiggins 1976: 146, 1980: 166.

36 Wiggins 1980: 166.

37 See Wiggins 1976 for this 'delta' terminology.

38 Wiggins 1980: 163ff.

39 Wiggins 1980: 169.

40 Wiggins 1980: 167.

41 Wiggins 1980: 166.

42 Wiggins 2001: 229.

43  Wiggins 1980: 166.
44  Wiggins 1980: 167.
45  Wiggins 1980: 167.
46  In slightly less august terms than Sartre's, we find a similar use of 'person' in the trope that pervades Western literature, of one's *becoming* or *being* one's parent (e.g. Gaiman 2013: Chapter 1, Goodison 1986: 'I am becoming my mother'). Describing the inheritance of character, Julien Green captures the common thought in his novel *Léviathan* (Green 1929/1993: 32): 'In growing older we become our parents.'
47  Wiggins 1980: 167.
48  Mauss 1938/1985: 15.
49  Mauss 1938/1985: 15–17.
50  Mauss identifies an interesting point of transition in the dissolution of the father's power over the life and death of his descendants (*vitae necisque potestas*). Prior to that dissolution, his children, being bearers of his name, were 'part of his person' (in the fullest sense) and he could do with them as he pleased (Mauss 1938/1985: 16).
51  Mauss 1938/1985: 16.
52  Mauss 1938/1985: 17.
53  Mauss 1938/1985: 17.
54  Thus stands the entry in the *Oxford English Dictionary* (available online at www. oed.com).
55  Consider too that a corporate person may be constituted by numerous and diverse individuals at different times, and can exist spatially dislocated (in different 'branches'). Perhaps this will provoke similar difficulties to the mask conception.
56  Bakhurst 2005: 463.
57  Snowdon 1996: 38.
58  Snowdon 1996:

> We would not countenance the possibility that the elements on the standard list elucidating person might be eliminated given investigation of a wider sample, nor that features with no obvious connection to those already on it might, on investigation, merit being put on the list.
>
> (38)

59  At times, he seems tempted to endorse such a reading (e.g. 1980: 173).
60  Wiggins 1987: 63.
61  Wiggins 1980: 180–181.
62  Wiggins 2005: 474.
63  Wiggins 2001: xi (and 242ff).
64  Bakhurst 2005: 463.
65  Bakhurst 2005: 464, fn. 6.
66  Bakhurst 2005: 464.
67  Mauss 1938/1985: 18–19.
68  Mauss 1938/1985: 18. Hollis presents what seems to me to be a fair summary of the development up to this point (Hollis 1985: 219).
69  Mauss 1938/1985: 19.
70  Hollis 1985: 217.
71  La Fontaine 1985: 123.
72  Nor, it seems, is Wiggins fully blind to this concern. He writes:

> In the classical era, Greeks and Romans seem to have found no need whatever for a term with the general sense or function of the word 'person'.[27] [The Latin word *persona* means primarily (1) mask or (2) a character in a drama. What first pushed the word in the direction of its modern meaning was the task of codifying Roman Law.]
>
> (Wiggins 1996: 248)

73   Mauss 1938/1985: 19.
74   Mauss 1938/1985: 17. See also La Fontaine:

> [the conception of *prosopon*] is translated by Christianity into the idea of the soul to arrive finally at the notion of a unity, of body and soul, mind and conscience, thought and action which is summed up in the concept of the individual which Mauss labeled '*la personne morale*'.
>
> (1985: 124)

See also Hollis 1985: 217.
75   Mauss 1938/1985: 20.
76   More generally, one might wonder whether the *person* concept can be reduced in this way, whether attempting to separate certain elements will damage that rich and complex notion. We may well think that the ambiguities between these different, occasionally conflicting elements, are an important feature of our living language and our way of interacting with others. This is prompted by Wittgensteinian thoughts conveyed to me by Christoph Schuringa.
77   See, for example, Wiggins 1980: 222: 'We entertain the idea, unless we are irredeemably conceited or colonialist in mentality, that there be something we ourselves can learn from strangers about the true, the good, and the rational'.
78   Wiggins 1995: 248.
79   Wiggins 1987: 63.
80   And indeed, attending to his comments in his 1987 paper, Wiggins does not seem to want to use the Strawsonian point to do anything other than gesture towards the connection.
81   Wiggins 2000: 1.
82   Note, too, how commonplace it is to wonder what kind of person a baby, as yet unborn or newly-born, will become. Looking at an ultra-sound image of a foetus we may well think the entity depicted there is not yet a person, but will one day be one.
83   Bakhurst 2005.
84   Bakhurst 2005: 467.
85   In anticipation of the discussions in Chapter 7 it should be noted that this picture fits neatly with John Dupré's (e.g. Dupré 2010/2012).
86   Wiggins 2005: 475.
87   Wiggins 1976, 1987.
88   Wiggins 1987: 63.
89   Wiggins 2000: 1. Note, too, that this connects to the moral objection to the nominal-essentialist account of 'person' (described above).
90   See also Wiggins 2001: 198.
91   Wiggins 2000: 1.
92   Wiggins 2000: 199: 'the marks of personhood are assembled of persons *as we know them* from the only case we shall ever become familiar with, namely that of persons who are human beings.'
93   For a helpful introduction to this Aristotelian thought see Boyle's 'Essentially Rational Animals' (*forthcoming*).
94   Wiggins 1967: 46–47.
95   Wiggins 1987: 72.
96   Wiggins 2001: 194.
97   In fact, attention to Wiggins's own various discussions of the relation between the concepts *person* and *human being* suggests an asymmetric dependence, characteristic of the dependence of a restricted sortal on a substance sortal. See Wiggins 1987: 60 and 75.
98   See, for example, Mauss 1938/1985: 8–9.
99   See, for example, Tur 1987: 117 (and Mauss 1938/1985: 17).

100  Locke (1690/1975): II, xxvii, §18. Note the descriptivist appeals in this passage; it is a commonplace of our natural way of talking to describe ourselves as *becoming* different persons. This is suggested too, by common phrases like 'she was a different person then' and 'you're a different person here than at work' (and so on).

## Bibliography

Allen, N.J. (1985) 'The Category of the Person: A Reading of Mauss's Last Essay', in M. Carrithers, S. Collins and S. Lukes (eds) *The Category of the Person: Anthropology, Philosophy, History* (Cambridge: Cambridge University Press).

Bakhurst, D. (2005) 'Wiggins on Persons and Human Nature', *Philosophy and Phenomenological Research* 71(2): 462–469.

Boniolo, G. (2013) 'Is an Account of Identity Necessary for Bioethics? What Post-Genomic Biomedicine Can Teach Us', *Studies in History and Philosophy of Biological and Biomedical Sciences* 44: 401–411.

Boyle, M. (*forthcoming*) 'Essentially Rational Animals', in G. Abel and J. Conant (eds) *Rethinking Epistemology* (Berlin: DeGruyter).

Craig, E. (1990) *Knowledge and the State of Nature: An Essay in Conceptual Synthesis* (Oxford: Oxford University Press).

Dupré, J. (2010/2012) 'Causality and Human Nature in the Social Sciences', in J. Dupré *Processes of Life* (Oxford: Oxford University Press).

Foucault, M. (1994) 'Nietzsche, la généalogie, l'histoire', in M. Foucault *Dits et Écrits*, vol. I (Paris: Gallimard).

Fricker, M. (2007) *Epistemic Injustice* (Oxford: Oxford University Press).

Gaiman, N. (2013) *The Ocean at the End of the Lane* (New York: William Morrow and Company).

Geuss, R. (1999) *Morality, Culture, and History: Essays on German Philosophy* (Cambridge: Cambridge University Press).

Geuss, R. (2001) *Public Goods, Private Goods* (Princeton: Princeton University Press).

Goodison, L. (1986) *I Am Becoming My Mother* (London: New Beacon Books).

Green, J. (1929/1993) *Léviathan* (published in English as *The Dark Journey*) (Quartet Books).

Hill, R.K. (1998) 'Genealogy', in E. Craig (ed.) *Routledge Encyclopedia of Philosophy* (London: Routledge). Available online at www.rep.routledge.com/article/DE024 SECT1.

Hollis, M. (1985) 'Of Masks and Men', in M. Carrithers, S. Collins and S. Lukes (eds) *The Category of the Person: Anthropology, Philosophy, History* (Cambridge: Cambridge University Press).

Koopman, C. (2012) *Genealogy as Critique: Foucault and the Problems of Modernity* (New York: Fordham University Press).

La Fontaine, J.S. (1985) 'Person and Individual: Some Anthropological Reflections', in M. Carrithers, S. Collins and S. Lukes (eds) *The Category of the Person: Anthropology, Philosophy, History* (Cambridge: Cambridge University Press).

Locke, J. (1690/1975) *An Essay Concerning Human Understanding*, P.H. Nidditch (ed.) (Oxford: Oxford University Press).

Mauss, M. (1938/1985) 'A Category of the Human Mind: The Notion of Person; The Notion of Self', in M. Carrithers, S. Collins, and S. Lukes (eds) *The Category of the Person: Anthropology, Philosophy, History* (Cambridge: Cambridge University Press).

Morton, A. (1990) 'Why There Is No Concept of a Person', in C. Gill (ed.) *The Person and the Human Mind: Issues in Ancient and Modern Philosophy* (Oxford: Clarendon Press).

Nietzsche, F. (1887/1996) *On the Genealogy of Morality* (Oxford: Oxford World Classics).

Rorty, A.O. (1990) 'Persons and *Personae*', in C. Gill (ed.) *The Person and the Human Mind: Issues in Ancient and Modern Philosophy* (Oxford: Clarendon Press).

Skinner, Q. (2002) *Visions of Politics: Vol. 1: Regarding Method* (Cambridge: Cambridge University Press).

Snowdon, P. (1996) 'Persons and Personal Identity', in S. Lovibond and S. Williams (eds) *Essays for David Wiggins: Identity, Truth and Value* (Oxford: Blackwell Publishing).

Tur, R. (1987) 'The "Person" in Law', in A. Peaocke and G. Gillett (eds) *Persons and Personality* (Oxford: Blackwell).

Wiggins, D. (1967) *Identity and Spatio-Temporal Continuity* (Oxford: Blackwell).

Wiggins, D. (1976) 'Locke, Butler and the Stream of Consciousness: And Men as a Natural Kind', *Philosophy* 51: 131–158.

Wiggins, D. (1980) *Sameness and Substance* (Cambridge, MA: Harvard University Press).

Wiggins, D. (1987) 'The Person as Object of Science, as Subject of Experience, and as Locus of Value', in A. Peaocke and G. Gillett (eds) *Persons and Personality* (Oxford: Blackwell).

Wiggins, D. (1995) 'Substance', in A.C. Grayling (ed.) *Philosophy 1: A Guide Through the Subject* (Oxford: Oxford University Press).

Wiggins, D. (1996) 'Replies', in S. Lovibond and S. Williams (eds) *Essays for David Wiggins: Identity, Truth and Value* (Blackwell Publishing).

Wiggins, D. (2000) 'Sameness, Substance and the Human Animal', *The Philosophers' Magazine*.

Wiggins, D. (2001) *Sameness and Substance Renewed* (Cambridge: Cambridge University Press).

Wiggins, D. (2005) 'Reply to Bakhurst', *Philosophy and Phenomenological Research* 17(2): 442–448.

Wiggins, D. (2016) *Twelve Essays* (Oxford: Oxford University Press).

Williams, B. (2002) *Truth and Truthfulness: An Essay in Genealogy* (Princeton: Princeton University Press).

# 6 Biological models

Leaving aside issues about the consilience of *person* and *human being*, let us fix our attention on the latter of the two concepts. Irrespective of the fate of *person*, Wiggins takes human beings to be *natural substances* – and examining his analysis one becomes aware of how his metaphysical project intersects with certain concerns in the philosophy of biology. Specifically, his work bears on, and is affected by, discussions about *biological individuality* and *biological anti-reductionism*. It is the aim of this and the next three chapters to draw out these connections.

It has already been noted that Wiggins enters forceful claims about the relevance of the philosophy of biology to metaphysical issues[1] – yet despite these claims, and occasional allusions to differing conceptions of that science,[2] he never offers a sustained treatment of these issues. Moreover, his view appears to verge on what he calls 'the anti-scientistic'.[3] Sometimes he denies his aim has ever been 'to place the question of what we are under the alien direction of physiologists, biologists, evolutionists or others who are expert in matters relating to *organisms*'.[4] Thus, commentators tend not to discuss the role of biology in his work, unless it is to bemoan his failure to attend to it. (Michael Ruse, for example, describes him as displaying 'an almost proud ignorance of the organic world'.[5]) This, it will be argued, is an oversight on the part of his critics. The following reading presents Wiggins as having a determinate view of the place of biology in his philosophical system – and these sections aim to collect together and interpret his thoughts on the application of that science in metaphysics.

There is one immediate point of connection between biology and metaphysics: Wiggins defers to biology in order to flesh out the *human being principle*, the theoretical description that allows us to articulate the spatio-temporal boundaries of human organisms. Within the sphere of personal identity, this has always seemed a relatively uncontroversial aspect of his approach. Nor is the assumption that biology sheds light on our persistence conditions to be found only in Wiggins's work. It is one of the strengths of the animalist position that their central theoretical posit – the 'human animal' – is taken to be a subject of study outside the rarefied environs of Analytic metaphysics. Olson, van Inwagen and others routinely direct us to biologists for a precisification of their accounts of human survival.[6] Yet, turning to the philosophy of biology, one finds lively debate about the possibility of articulating such a principle. It is this problem that we will consider first.

## 1  Limits of living activity

Wiggins states that when we pick out and track human beings – as we do routinely in our daily lives – our efforts are guided by a clear, if indistinct,[7] idea of what the *principle of activity* of a human being is. We have a rough and ready conception of the activity of such a thing and, he writes, *biological inquiry* can tell us more about the typical growth patterns, behaviour, development, etc., of creatures like us:

> [W]e are led by simple conceptual considerations to precisely the account of living substances that biologists can fill out *a posteriori* by treating them as systems open to their surroundings, but so constituted that a delicate self-regulating balance of serially linked enzymatic degradative and synthesizing chemical reactions enables them to renew themselves on the molecular level at the expense of those surroundings, such renewal taking place under a law-determined variety of conditions in a determinate pattern of growth and development towards and/or persistence in some particular form.[8]

However, when we turn to biology to precisify our understanding of these everyday individuals, certain puzzles start appearing. The first is one much remarked upon in the philosophy of biology literature: the notion of an essential 'activity' assigned by some 'theoretical description' common to a single species is deeply problematic in the light of the various critiques of species essentialism.[9] Given his association with Kripke and Putnam, Wiggins appears at points to be advocating this form of essentialism. This concern will be discussed at the end of this chapter in relation to an objection from Samir Okasha. But in general the intention here is to consider another, more marginal (though no less important) puzzle: how in the first place are we to fix on the biological individuals among whom the theoretical description is supposed to hold?

How do we define the target of biological inquiry? Where do we find the site of the living activity whose functions we wish to scrutinise? Picking human organisms out 'phenomenally' seems relatively straightforward – we just 'see' organisms and take them to be delimited, for example, by the boundaries of their skin. (Here, 'phenomenally' is used by philosophers of biology to describe the singling out of entities that happens in our day-to-day lives – the more technical, philosophical connotations of the word are suppressed, though the lien between the two will become evident (to Wiggins's advantage).) We pick out cats, dogs, trees, flowers and other humans, and, on the face of it, we do so with relative ease; it is *this* thing here (you point to yourself) whose activity we wish to examine.

Probing a little deeper, however, we find matters are not quite so simple. There are things within the bounds of your skin whose to-ing and fro-ing we would not want to include as aspects of your living activity. You do not – I hope – have a tapeworm nestled in your gut, but finding one we would not want to include its parasitic endeavours as part of your life. The same would be true if we were to discover a tumour. There is a disturbingly large number of entities that can lodge themselves inside human organs and take up residence. They do not contribute to one's life but rather drain it. The 'natural boundary' of the skin is not so good a marker of human limits as we might at first assume.

There are, on the other hand, creatures that inhabit human bodies which are neither harmful nor accidental, who might actually be seen to enhance or, to some degree, *sustain* your or my living activity. Consider, for instance, the bounty of gut flora that line your insides. As philosophers of biology are often wont to mention, investigation at the microscopic level shows that human organisms can be seen, as John Dupré puts it, as 'symbiotic system[s] containing [multitudes] of microbial cells – bacteria, archaea, and fungi – without which the whole would be seriously dysfunctional and ultimately non-viable'.[10] We enjoy numerous, mutually beneficial relationships with the vast numbers of endosymbionts that live within us, whose collaboration makes our lives possible: firmicutes, bacteriodetes, actinobacteria and proteobacteria, creatures whose austere names belie the intimacy of our relation to them. Are these endosymbionts parts of us? The processes they are involved in – digestive, immunological, etc. – are not unimportant for our survival.[11] Do we include their activities in the 'principle of activity' of the human being? This question has risen to increasing prominence in the light of work by Dupré and Thomas Pradeu. Jack Wilson is one who has suggested how the problems it engenders might reach into the sphere of 'personal identity'.

The difficulty becomes even starker when we consider other types of living being. Wiggins's account of individuation is intended to be general. But we are constantly discovering facts about other 'phenomenally' discrete entities which throw our everyday opinions into disarray. Wilson offers us the example of the *nanomia cara*.[12] The *nanomia cara* is an aquatic creature that to all intents and purposes looks and acts just like a jellyfish. But while jellyfish are strange enough (placed, by Aristotle, in a liminal realm between plant and animal kingdoms), the *nanomia cara* is stranger still – despite appearances it is, in fact, a colony of various bacteria, working in the style of a corporation, to feed and move and grow together.[13] How do we delineate this organism's 'living activity'? Should we even call it 'an organism'? Even if they slip undetected passed the phenomenal, everyday gaze, conglomerates like these are in fact quite common.

And what of the plant world? Trees, like the aspen, and fungi, such as *Armillaria bulbosa*, are routinely picked out as individuals but they are often connected by vast networks of hidden roots to hundreds of other plants which are their genetic clones. Such trees might more properly be called proper parts of 'super-organisms'.[14] Plants and fungi are, when we attend to them, profoundly peculiar beings and on microscopic investigation they confound our everyday expectations of how 'organisms' behave. From the giant redwood to the smallest slime-mould[15] the bounds of living beings are confusing. 'The universe', as J.B.S. Haldane had it, 'is not only queerer than we suppose, but queerer than we *can* suppose';[16] scientific investigation routinely troubles our everyday, phenomenal modelling of the biological realm. Thus fixing on entities for scientific elaboration is nothing like as straightforward as Wiggins or the animalists assume.

Here, we are presented with a question which, like its subject, becomes more complex as we study it. How can we set a principled limit to an investigation into organismic activity? We must define the organism. But to do so we must have some notion of the activity that binds its parts into a unified whole. Is there a theory that can offer us surer foundations for picking out 'phenomenal' individuals and understanding their lives? Such questions – side-lined in discussions of personal identity[17] – are raised by these philosophers of biology interested in *biological individuality*.[18] Although we seem relatively unproblematically to be able to pick out human beings (and other animals), this phenomenal ability of ours falls short of determining what counts as a *part* of an individual. Are we to include or exclude the glut of gut flora that enhance our lives, the endosymbionts without whom we would experience considerable – sometimes fatal – difficulties? In this chapter I shall set out some of the different responses to this concern. *Functional*, *genetic*, *epigenetic*, *autonomy* and *immunological* accounts offer ways of sharpening our biological focus. All of them pick out 'organisms', but they do so differently. Which of them align best with our pre-theoretical framing? Where they do align, can they be used to justify our everyday partitioning of the biological world – or are they somehow dependent upon it? In examining these issues, we will be led to another concern, to be expounded in the next chapter. Scientifically precise notion of a living being may conflict to some degree with our pre-theoretical partitioning of reality.

### (i)  *The physiological-functional account*

The picture most immediately amenable to Wiggins's aims seems to be the 'physiological' or 'functional' account. 'Physiology', here, is understood to denote the biological fields that focus on 'how' questions: *how* does the blood flow around the body, *how* does the liver work, etc. – i.e. those fields, like anatomy and morphology, which aim to understand an entity's *functioning*. Thus Ernst Mayr refers to these fields as instances of 'functional biology'. I shall call this account the 'physiological-functional' one. This view, advanced by those like Eliot Sober and H.P. Wolvekamp,[19] claims, roughly, that an organism is a 'functionally integrated' unit, made up of causally interconnected elements.[20] As Stephen J. Gould states the position, a biological item

> is an organism if it is spatially separated from others and if its parts are so well integrated that they work only in co-ordination with others and for the proper function of the whole.[21]

The thought is an intuitive one. We might find, for instance, a woodlouse, firmly lodged in the bark of a tree. These things – the louse and the tree – are very close, spatially contiguous even, but they are not functionally integrated. In a disruptive spirit, we might dislodge the louse from his home and he would find another. The tree would suffer no form of impairment. Contrast this with the sadistic subtraction of parts from the louse (an organ, say). This he would notice. The woodlouse is composed of heterogeneous causally integrated parts. He is a causal unity, whose ability to function would be disrupted – perhaps fatally – by the loss of one or other of those parts. The physiological-functional account builds on this conception and seeks on this basis to articulate the boundaries of the functional system.

The intuitiveness of this picture is of course one of its strengths. As Sober notes, it has the benefit of aligning with common-sense; it seems clear to us that our livers and hearts – and not our dentures or scarves or malicious parasites, if we have them – are *proper* parts of us because their activities are deeply bound up with the other processes that go on inside us.[22] The physiological-functional view appears therefore to answer the concern voiced above: when we pick out human beings, we pick out functional units. In this way we can determine the limits to the target of inquiry (into principles of activity) by considering where the limits of the *functional* unit lie.

There are two immediate problems with this approach. First, as Wilson points out, functional integration clearly comes in degrees. 'The organs of a mouse are causally integrated; [but] so are the members of a good pit crew or state legislature.'[23] The line, then, becomes hard to draw. We might exclude dentures as functional parts while including other, more advanced 'prostheses' (artificial hearts, for instance). As it stands, 'functional integration' is too vague. At what point are things functionally integrated? How functionally integrated are the foetus and its mother? The physiological-functional view fails to offer us a principle for including or excluding endosymbionts in the biological 'unit'. It remains, as Pradeu notes, awkwardly gestural: 'we simply trust our impression that the organism is a coherent "whole", which we cut into functional pieces'.[24] And it is this phenomenal 'impression', which is precisely the picture that is under investigation. Any defence of it made via this route would seem to be circular.

There is also a second-level concern. As Wilson points out, 'the components of a single cell are well integrated, yet that cell may be a part of a multicellular organism in which all of the cells are also integrated into a collective functional individual'.[25] That is to say, similar degrees of functional integration occur at different levels; cells are spatiotemporally localized and functionally integrated, but so are symbionts, and 'colonial' organisms, which have, for example, common vascular networks.[26] There are innumerable overlaps. Should we somehow fix on a base-level for functional integration – excluding, for example, metal hips but including gut flora – we would still have to explain the relation between the larger functional units and the smaller functional units that constitute them.

The result is that, as it stands, the 'physiological-functional integration' account is not enough to provide theoretical grounds for demarcating organismic limits. It corresponds to the phenomenal picture, but it does so by relying on it. It cannot stand to justify or explain it. These at least are the common complaints about this model.[27]

## (ii) The genetic account

Perhaps, instead, we should make reference to the 'human genome' when talking about what is and what is not a part of the human being, in order to determine the limits of the living activity. On the genetic view, which one finds in the work of Maynard Smith and Richard Dawkins,[28] the organism's parts are conceived as genetically homogenous. Each cell in our bodies has a characteristic genetic make-up which is unique and distinctive to humans. Symbionts, then, are excluded as parts because they are genetically different.[29] It is through mapping our genetic codes that we can better understand our distinctive principle of activity. Unlike the gestural physiological-functional account, the genetic picture offers a principle, sustained by a sophisticated theory, by which to demarcate the focus of inquiry. The theory in question is the *theory of evolution by natural selection*; the individual is not picked out according to functional or causal unity but rather on the basis of whether or not it figures as *a unit of natural selection*. The limits are the limits of the *evolutionary* individual, i.e. that which is the subject of evolutionary processes.[30]

However, despite a pervasive assumption that there is a one-to-one relation between an organism and its genome, philosophers of biology now see this idea as deeply problematic. One relevant worry is that to think in this way legislates, as Pradeu puts its, a radical 'revision of our ontology'.[31] The genetic view constitutes a significant departure from our everyday world-view. Consider, for example, the case of monozygotic twins.[32] They share the same genotype (arguably), and would thus register as a *single* biological individual. The same is true for aphid colonies (which can be the product of asexual reproduction); they are genetically homogenous, so the individual things we typically pick out (and try not to squash) are configured as parts of a larger biological unit.[33] As Dupré makes clear, in his paper 'The Polygenomic Organism', genetic individuals can be composed of spatially discontinuous clones, so the genetic view fails to map onto our pre-theoretical phenomenal picture. (This distinction can be construed in Harper's terms as the diverging images of the *genetic* and *rametic* individuals.[34])

Following Wilson, we can say that the genetic individual has dramatically different persistence conditions from the organisms of our everyday understanding. It starts as a single cell,[35] produced from the fusion of genetically distinct gametes. It can then grow to the size of a human – or maybe two, if they are monozygotic twins – and it can be reduced back down to the size of a single cell. This is not a reduction that we think organisms can undergo and survive.

In short, while the genetic account has a theoretical support that the physiological-functional account lacks, drawing organismic boundaries along *genetic* lines will lead us to lose sight of the entities we pick out phenomenally. This is a price some metaphysicians might be willing to pay – yet, irrespective of its internal coherence, the genetic account does not appear to provide the kind of detailed biological framework that is needed.

It is possible that some pairing of the genetic account with the physiological-functional account might capture the phenomenal individual. This, ultimately, is the position I take Wilson to be advocating in *Biological Individuality*.[36] We can say that the human being is a functional individual, whose parts are included as parts based on their genetic homogeneity. The physiological-functional element would allow us to distinguish between, for example, monozygotic twins and the genetic element would provide a principled way of demarcating organismic limits. Nevertheless, there are philosophers of biology – among them, Dupré and Pradeu – who would disagree strongly with Wilson's thought that 'the functional individual has a unique genetic constitution'.[37] Even if we ignore the problematic cases of endosymbionts, the pervasiveness of genetic mosaicism and chimerism undermines this claim.[38] As in tortoiseshell or calico cats, parts of the same functional individual can have different genomes. Dupré lists a variety of naturally occurring and artificially created chimeras, which fulfil Wilson's criteria for functional integration without being genetically homogenous. As things currently stand, neither the physiological-functional account nor the genetic account fit the bill.

### (iii) The epigenetic account

The epigenetic account, presented by Giovanni Boniolo and Giuseppe Testa, emerges in no small part in response to the perceived narrowness of the genetic approach.[39] In particular, Boniolo and Testa seek to address the problematic assumption of *genetic essentialism* which, until relatively recently, pervaded the literature both inside and outside the philosophy of biology. (As we shall see, Okasha finds it seeping into Kripke and Putnam's work on natural kinds.[40]) The genetic essentialist holds that the living process that sustains an organism is simply the output of a genetic 'blueprint'. There is a causal chain, from the genetic level to, for example, the morphological level, which goes one way and one way only. This essentialist position conflicts, however, with a considerable amount of 'epigenetic' research, the area of research that focuses on 'external' factors that organize gene transcription. Dupré, again, is one who has repeatedly emphasized the significant roles that epigenetic processes play in animal development, often using 'methylation' as an example (something we return to in Chapter 7):

> [T]he genome itself is modified during development, a process studied under the rubric of epigenetics or epigenomics. The best known such modification is methylation, in which a cytosine molecule in the DNA sequence is converted to 5-methylcytosine, a small chemical addition to one of the

nucleotides, or bases, that make up the DNA molecule. This has the effect of blocking transcription of the DNA sequence at particular sites in the genome. Other epigenetic modifications affect the protein core, or histones, which form part of the structure of the chromosome, and also influence whether particular lengths of DNA are transcribed into RNA.[41]

Living processes are not simply the results of genetic instructions; rather, they figure here as dialogues between genes and their environment. Oftentimes, changes in the make-up of DNA occur, during one life-cycle, in response to higher-level factors. This organic 'dialogue' is the focus of the account endorsed by Boniolo and Testa, who think that epigenetics will provide a firmer foundation than the genetic picture, for answering what they call the 'Who-problem': 'what are the biological properties that make a living being unique and different from others?'[42] Their answer is that

> the living being is its whole phenotype, intended as the result of the epigenetic processes that have regulated the expression of its genome up to that time-slice.[43]

Or, again,

> A human being, in any instant of his/her life, is nothing but the result of all the epigenetic processes that, in the course of time, have causally shaped all of his/her interconnected phenotypic modules.[44]

Unlike the genetic account, their model does not restrict the living activity to those processes that issue from exclusively genetic directives. They avoid the trap of genetic essentialism. Moreover, by emphasizing the importance of environmental factors, they appear to have a principle for distinguishing between clones; identical twins can be considered distinct individuals because, being spatially discontinuous, they are subject to different epigenetic influences.[45]

The epigenetic account certainly seems to make up for some of the deficiencies of the genetic account. But does it better capture Wiggins's phenomenal/pre-theoretical individual? Consider again Boniolo and Testa's statement of what a living being is: '*its* whole phenotype'.[46] *Whose* whole phenotype? The living being is the living being's whole phenotype. On the one hand this appears recursive; on the other hand, it suggests that Boniolo and Testa already have in mind a rough idea of the living individual, on which they wish to sharpen their focus. On a positive reading, they see – like Wiggins – living beings as easily individuated, and the aim of their epigenetic approach is to strengthen our conceptual grip on them. To some degree, this is supported by the terms of their distinction between the living being and its 'environment', which their position presupposes. 'For a living being, environment starts from the outer layer of its body …';[47] i.e. the skin. This was precisely the 'natural boundary' that, at the start of this section, was identified as the limit typically registering in phenomenal individuation.

The problem is that on this (admittedly cursory) reading their approach seems to do nothing to *substantiate* the phenomenal boundaries. The living being that we fix on in our everyday lives is the starting point for their investigation. Its limits are neither explained nor scrutinized. Thus, the questions with which we began are not really answered. We are not told, for example, whether the endo-symbionts within us are parts of us. On the basis of the quotation above, the gut flora that line our insides cannot be considered as environmental entities; they are *assumed* to be elements of our living activity. We are offered no principle of inclusion for this, other than that they reside within our skin.[48] Thus, while the epigenetic account may fare better than the genetic picture, it relies too heavily – one might think – on the phenomenal account. Like the physiological-functional view, it cannot be seen to substantiate or confirm our everyday partitioning of cats and dogs and humans. Nor are we offered theoretically precise ways of investigating their principles of activity.

### (iv)  The autonomy account

Where else might Wiggins look? He might do worse than turn to the kind of 'autonomy' account advanced by those like Alvaro Moreno, Matteo Mossio, Kepa Ruiz-Mirazo and Juli Peretó. These theorists fully recognize the problem of biological individuality. As Ruiz-Mirazo and Moreno put it, 'evolution … weaves a collective network of increasingly complex and entangled cooperative relations among entities at different phenomenological levels and with different cohesive strengths'.[49] The problem is how to separate out individuals at these different levels, the endosymbionts from their hosts or partners.

Mirazo and Moreno's answer rests on the thought that living systems are organizationally *closed*.[50] Like Gould and Wolvekamp, these members of the 'Basque group' see part-hood to depend on functional integration within that closed system.[51] But their picture goes beyond the functional one in deferring to the notion of 'autonomy'.[52] The organizational system is an 'autonomous' agent, where 'autonomy' is for Ruiz-Mirazo and Moreno,

> the property of a system that builds and actively maintains the rules that define itself, as well as the way it behaves in the world. So autonomy covers the main properties shown by any living system at the individual level: (i) self-construction (i.e., the fact that life is continuously building, through cellular *metabolisms*, the components which are directly responsible for its behaviour) and (ii) functional action on and through the environment (i.e., the fact that organisms are *agents*, because they necessarily modify their boundary conditions in order to ensure their own maintenance as far-from-equilibrium, dissipative systems).[53]

In response to a changing environment, and in order to ensure upkeep and survival, living systems must build, maintain and modify certain boundary conditions.[54] For the Basque group – if not all 'autonomy' theorists – the mechanisms that deal with boundary drawing necessarily serve to demarcate biological individuals: 'an autonomous system is an organisation of functional relationships among distinguishable components whose causal effects is their cohesive or cooperative integration within that self-maintaining and self-producing organisation'.[55]

To function in the way they do, as relatively stable systems amid the constantly fluctuating storm of the biological realm, living beings must possess some mechanism for regulating that stability. The focus here is on the *metabolic* system – 'a cyclic, self-maintaining network of reactions by means of which the components of a system are continually produced in far-from-equilibrium conditions'[56] – and its *homeostatic* elements which serve as a guide to organic limits. It is not insignificant that biochemists think of metabolisms as 'enzymatically closed', i.e. as a complex network of chemical reactions controlled by enzymes that are themselves fabricated within the network. It is possible, Ruiz-Mirazo and Moreno think, that a mechanism like this can provide a theoretically refined basis for stating what falls inside and what falls outside a biological system.

Again, the overview is cursory, but the notion of *autonomy* might allow one to clarify the limits of the everyday organisms by deferring to the boundary-defining mechanisms, like metabolic processes.[57] As an approach to individuality it is scientifically more informed than the physiological-functional account. At the same time it is less severe than the genetic account and it corresponds more readily to our everyday thoughts about organisms. (Members of the same clone do not share a metabolic system, so they can be seen to be biologically distinct.) Nor is this account as reliant as the epigenetic account upon phenomenal individuation. The boundary of the skin is not taken to be a sufficiently clear principle of inclusion. Furthermore, by deferring to metabolic cohesion and self-maintenance, the autonomy theorists have the means to draw theoretically justified distinctions between autonomous conglomerations of cells and higher-order autonomous entities. (Consider here the *nanomia cara* and the *grex*.[58]) Their picture is grounded in our everyday conception of the world and deploys a principle of inclusion that seems to refine rather than to overhaul it.

As ever, however, there are reasons for doubt. For example, Ruiz-Mirazo and Moreno point out that their project encounters difficulties when applied to plants and fungi. Plants and fungi do not, they say, realize the strong metabolic, integrative cohesion that we find in multi-cellular metazoans. Thus, 'the decision to consider many of these multicellular systems as fully-fledged individual organisms is a matter of degree rather than of clear conceptual differences'.[59] The difficulty is that this vagueness does not, in the end, correspond with phenomenal individuation. Digging up a daffodil, I tend to think of it as a discrete individual, which can be re-identified at a later stage, possessing determinate boundaries. When I eat a mushroom I do not think I am eating a part of a larger entity. (Nor is this the only concern. The puzzle with which we started was whether endo-symbiotic activity should be included with an entity's living activity – and it is not obvious the autonomy account answers this.) When more is said about the principles of inclusion as realized by the metabolic system this worry might be overturned – but there is work still to be done here.

### (v) The immunological account

If, like me, Wiggins finds the autonomy project slightly obscure, he might fare better with Thomas Pradeu and Edgardo Carosella's *physiological-immunological* view.[60] I noted that our phenomenal framing corresponds to the physiological-functional account endorsed by Sober. I also remarked that Sober's view rested on too vague a notion of 'functional integration'[61] and relied too heavily on the phenomenal picture it was required to explain.[62] In that it aims to ground functional individuation in biological mechanism that sets boundary limits to a system, Pradeu's picture pursues a similar tactic to the autonomy account (and to Wilson's genetic-functional hybrid).

Pradeu has presented a way of supplementing the standard physiological-functional account with a 'criterion of immunogenicity'. His position is clearer to me than the autonomy theorists, and the principle of inclusion that he describes is more determinate (thus meeting the complaint mentioned at the end of the last sub-section). Pradeu claims that the *immunological system* operates as a principle of organismic inclusion, and can accordingly be used to draw the relevant lines to demarcate the organism. In his influential paper 'What is an organism?' he writes:

> [T]he immune system, by its surveillance activity, defines what will be accepted, and what will be rejected, by the organism, and therefore a criterion of immunogenicity constitutes a *criterion of inclusion* for the organism: the distinction between the entities which will stick together as constituents of the organism, and those which will be rejected from the organism, is made by the immune system.[63]

Thus the immune system provides principled grounds for drawing the boundaries of the functionally integrated 'whole'. It is a sub-system, the activity of which determines what is and what is not a proper part of an organism. It is, to my mind, more precise than the self-maintaining mechanisms that the Basque group defers to. Moreover, this account indicates a primary level of functional integration. Cells may well be functionally integrated, but the immune system provides grounds for defining them as parts of a greater unity.[64] Pradeu's focus on the immunological allows him to bolster the standard physiological-functional account and find determinacy where previously there was vagueness.

The physiological-functional account was seen to correspond to our common-sense individuation. We pick organisms out as functional units. If successful, this view can provide a means by which to refine this picture. Following Pradeu, Wiggins might find a sound scientific basis by which to identify certain interconnecting biological interactions[65] – those moderated by the immune system. He would say that these are the physiological processes that constitute the 'principle of activity' of the everyday organism.

Let us mention, just briefly, one worry that troubles this picture. Pradeu notes that his approach construes organisms as *heterogenous entities*. In contrast to previous immunological accounts[66] he departs from what is known as the 'self-nonself' criterion of immunogenicity, which claims that an organism does not trigger immune responses to its own constituents but to *every foreign entity*. Pradeu's theoretical shift accommodates the immune role that *exogenous* bacteria play in our bodies (they register, immunologically, as 'parts' of the body). This nicely illustrates the complexity and symbiotic involvement in living process. The worry is that his account construes organisms as having heterogeneous constituents; you are not just your cells, you are also all of those beneficial bacteria that exist symbiotically within you.[67] Even Pradeu takes this to be 'counter-intuitive' to our ordinary thoughts about organisms.[68]

Should this detain us? Maybe it seems odd to say that revelations at this level of biological scrutiny are 'counter-intuitive'. We are experientially unaware of the huge numbers of beneficial bacteria that cover our skin and the lining of our guts so there will be no intuitions about whether or not they are parts of us. Might Wiggins tack thus and say that, on the phenomenal level, a human being conceived as an organism with heterogeneous constituents is not relevantly different from a human being seen otherwise?[69] Perhaps this finding is precisely the kind of information that biological inquiry into the human being principle was expected, hoped even, to deliver? I think there is more to be said here – and, after a partial summary and analysis, we will return to this problem of our minuscule, microbial body parts.

## 2  A simple-minded worry?

Though the list above is not exhaustive, it demonstrates the variety of theoretical models for picking out *biological individuals*. There are physiological-functional individuals, and genetic individuals, alongside immunological ones, and more (developmental, historical and epigenetic ...).[70] The question has been whether any of these pictures single out *organisms* and among them *human beings* who are the immediate subject of our everyday attention. This question is pressing for Human Being theorists and animalists alike. If they are to defer to biological models to support their views, then they must first choose the models they will defer to.

The physiological-functional picture, with which we started, was instantly appealing. Ordinarily, we pick out living things as functional units. Organs appear as proper parts of animals because they are causally integrated in the right way. Parasites like tapeworms are not. Yet, while this view could help to extrapolate our everyday picture it relies too heavily upon it to substantiate it. By itself, it cannot tell us whether, for example, endosymbionts are parts of us. It turns us to our commonsensical impression 'that the organism is a coherent "whole"'.[71] Like a shadow, it follows in the footsteps of pre-theory, neither substantiating it, nor questioning it. The same problem was seen to hold for Boniolo's epigenetic view.

Subsequent accounts were seen to tackle this worry by combining the physiological-functional account with some other model. The genetic account does not by itself capture the entities of our everyday awareness[72] but those like Jack Wilson think that it can be used to *bolster* the functional picture. We can admit as parts only those bits that are genetically homogenous.

There are problems with Wilson's approach – his claim about genetic homogeneity, for instance, is belied by the pervasiveness of chimerism and mosaicism – and resultantly, the 'autonomy' theorists turn from the genetic model to a 'metabolic' principle of inclusion. Similarly, Pradeu attempts to shore up the physiological-functional account with an immunological element. There is clearly some mileage in these attempts. The individuals they describe are less emaciated than the genetic entity and seem to draw the conceptual net closer around the beings that hold our interest. This overview suggests that these accounts *might* offer Wiggins a biologically precise way of picking out human organisms.

Yet a worry remains, one hinted at briefly in the analysis of Pradeu's account. Wiggins is interested in the commonsensical organism. The problem with the immunological view is that it describes us as *heterogeneous* beings made up of living bacteria – and that's just not how we think of ourselves. In some ways this is a simple-minded worry. In others it is one of considerable import which points to a fissure that grows between a 'scientific' world-view and our ordinary pre-theoretical framing.[73] With all the accounts above – excluding, notably, the physiological-functional one – there is a discrepancy between their theoretical posits and the human being, the subject of our moral, emotional concerns. Irrespective of any intuitions about bacteria, good or bad, we certainly *do* have intuitions about the kinds of entities that we are. We do not ordinarily think of ourselves as being made up of minuscule creatures. The claim, put bluntly (and developed below) is that we see ourselves to be *genuinely unified*, possessed of a single, discrete life, which is not shared among participants. This view of ourselves and of our kith is a central part of our pre-theoretical framework. We do not see each other as heterogenous entities. Indeed, our everyday eyes do not even countenance the individual cells or bacteria that make us up.[74]

These scientifically-informed accounts cannot correspond to the commonsensical picture because they posit (justifiably) entities that fail to register in our pre-theoretical framework (bacteria and genes, not least). That is, to repeat J.B.S. Haldane's now somewhat ominous phrase, the universe is queerer than we *can* suppose. In our ordinary lives we see each other in one way, overlooking the microscopic realm. The scientific view, by contrast, has to see it all; medium-sized objects have to be understood as heaving morasses of living specks. It is the work of the next two chapters to argue that to think of ourselves in this way is to think of ourselves as having a different metaphysical character to the organic *substances* of pre-theory.

So despite the diversity of biological models, Wiggins seems no nearer to finding a frame on which to hang his Human Being Theory. The account that came closest – Sober's physiological-functional view – was seen to be theoretically impoverished. 'Biologists have been engaged in the study of anatomy and physiology for centuries, but no "theories" of morphology and physiology have materialized.'[75] Yet the introduction of some such theory (like the immunological one) would disturb the intuitive aspect which attracted us in the first place. We would be forced to see ourselves as heterogeneous beings. Is there another way to substantiate Sober's model? I think there is an answer and, while it awaits the completion of the argument developed in Chapter 8, a provisional rendering can be offered now.

Wiggins's project is a descriptivist one – his metaphysics is guided by the structure of our conceptual framework and the entities that interest him are the entities we cannot help but attend to (like human beings). So the first thing to note is that, unlike the science-led metaphysician, Wiggins need not refer to science to divine which entities there *are*; the descriptivist can examine the metaphysical character of substances, whether or not they register in our 'best scientific theory' (Chapter 1 offered some arguments in defense of the metaphysical credibility of the descriptivist's project; Chapter 8 will further these thoughts with respect to *organic substances*). If, in the end, the physiological-functional view does not map onto some 'basic scientific laws' this need not adversely affect the descriptivist's use of it.

This tenders a second point: our phenomenal partitioning of the world is not as shallow as was first assumed. In responding to the problem about part-hood, the physiological-functional theorists were seen to trust their 'intuitions'. Then, examining things about which we have no intuitions – such as endosymbionts – they were flummoxed. Here, however, the descriptivist may lend a hand; she can examine the thoughts that underlie our intuitions. Some of them might be socio-culturally located, for sure, but others might be structural features of our conceptual framework. In Chapter 8 it will be argued that our pre-theoretical structuring of the world construes livings beings as *genuine unities*. It is suggested that this provides another approach to the problem of endosymbiont part-hood: symbionts are not proper parts of us, because, as per our pre-theoretical concept of *organic substance*, we are not heterogeneous entities. The physiological-functional picture can rely here on the phenomenal approach because, in the hands of the descriptivist, it can offer more than knee-jerk intuitions, but rather precise organismic boundaries (an advantage, note, of Wiggins's approach over the animalists').

This suggestion bears refinement, which it will receive in Chapters 7 and 8. The limits of the apparently drastic split between the 'scientific framework' and the 'pretheoretical' one will be considered in Chapter 9. For now, however, let it stand as a provisional response to the perplexities rehearsed here and turn our attention to another, related problem, which concerns Wiggins's apparent endorsement of species essentialism.

## 3 'Pluralism' and essentialism

While they might not correspond to our everyday *organism* concept, the models described above all seem to latch onto certain firm features of reality. We might not think of ourselves as genetic individuals, but this should not prejudice us against their existence or stop us recognizing that the theory that posits them has significant predictive and explanatory powers. Indeed, each of these models seems to pick out genuine entities with law-given limits. The recognition of this has in the philosophy of biology literature motivated a particularly liberal, 'pluralistic' attitude to ontology. We find it in Dupré (as we will see in Chapter 7), who argues for a form of 'promiscuous realism', where such entities have 'equal claims to reality'.[76]

Jack Wilson, in a similar spirit, construes a metaphysical liberalism of his own in terms of 'patterns', based on a notion of Daniel Dennett's.[77] Each of the above models of biological individuality recognizes a specific type of 'robust' pattern in nature, which 'identifies a group of phenomena that share important causal or law-like similarities'.[78]

> The existence of one pattern in a phenomenon does not preclude the existence of another pattern in the same phenomenon, even if each of these patterns is the basis of a natural kind.... The discovery of one natural kind of living entity, for example, the *genetic individual*, does not preclude the existence of a member of the natural kind *functional individual* even if functional individuals sometimes overlap with genetic individuals.... Patterns can overlap without impugning the reality of the overlapping patterns.[79]

This is a 'realism' of an open-minded sort. It does not try to accommodate *incompatible* entities. Nor does it attribute to a single entity conflicting persistence conditions.[80] The same is true for Dupré's position.[81] And both are notably similar to Wiggins's brand of verdant realism, set out above. (Indeed, both – Dupré (implicitly) and Wilson (explicitly) – seem to rely on Wiggins's notion of material coincidence in setting out their positions.) This connection suggests a way for Wiggins to respond to the accusation of 'genetic essentialism' mentioned at the start of this chapter.

The charge of 'genetic essentialism' comes from Samir Okasha (who, along with Wilson, deserves commendation as one of the few philosophers of biology to seriously engage with Wiggins's texts). There are various forms of essentialism – some of which, like 'sortal essentialism',[82] Wiggins clearly endorses. But, among these, 'genetic essentialism' is the most widely maligned. As described in Okasha's 'Darwinian Metaphysics: Species and the Question of Essentialism',[83] genetic essentialism is the thesis that members of the same natural kind share a genetic essence. In that essay, he states that Wiggins endorses Kripke's and Putnam's view that the 'theoretical descriptions', holding between exemplars of a kind, register at the genetic level.[84] He writes:

> Wiggins, like Kripke and Putnam, assumes without argument that organisms belong to their species in virtue of their "hidden structures" – their internal, presumably genetic properties.[85]

Like many philosophers of biology, Okasha argues that the distinction between species cannot, in the end, appear at any 'ultimate' genetic level.[86] We already have a rough idea of the reasons for this. The pervasiveness of chimerism, mosaicism and epigenetic processes disturbs the idea that an organism – let alone a *kind* – is genetically homogeneous. Let us accept Okasha's assigning of this sort of picture to Kripke and Putnam. Unlike Wiggins, Kripke and Putnam *do* suggest there is something like a genetic essence that underpins morphological similarities of natural kinds (Putnam states – of lemons – that the true criterion for membership in the kind *lemon* is having the 'genetic code' of a lemon;[87] Kripke talks of a shared 'internal structure';[88] Eric Olson falls with the same error[89]). Yet these views are not Wiggins's views. We will see in due course how he explicitly resists this reductive picture, but it should already be clear how his verdant realism recognizes theoretical descriptions registering at levels other than the genetic.

Okasha's reading might be based on claims in *S&S* about speciation (e.g. 'If we are interested in evolution and speciation ... then no doubt genetics (and ultimately molecular biology) must be at the root of what we inquire into'[90]). Such claims, however, are claims about 'species' where, in the context, 'species' is *not* synonymous with 'natural kind'. They do not commit Wiggins to an exclusively 'genetic' construal of the principle of activity of a kind. In *S&SR*, he draws a distinction between the principle of activity and the '*scientifically basic laws*'[91] – but in line with the interpretation offered above, being 'scientifically basic' need not correlate with being metaphysically fundamental.

For Wiggins, investigation into the biological realm can occur on different levels, and the examination of the human mode of being need not correlate with an examination of genetics.[92] The theoretical descriptions that hold between physiological-functional individuals will be different from those that hold between genetic individuals, or immunological ones, and so forth. Wiggins contends that there is a principle of activity for human beings – and this might be independently suspect.[93] But he does not need to hold that this activity is captured at the genetic level. Wiggins is not a genetic essentialist.

\*   \*   \*

Where do these perambulations leave us? We have articulated a problem for both the Human Being theorist and the animalist. It is not immediately obvious how best to flesh out the 'principle of activity' for human beings, because it is not clear how best to latch onto human beings. We catalogued a variety of methods for doing so, each with its own merits and failings. In comparing these models, a second question surfaced: can such scientifically informed theories hope to substantiate our pre-theoretical world-view? We are interested in human beings and we do not, ordinarily, see ourselves as the biologists see us, as made-up of constantly fleeting swarms of particles. A rough answer was offered in response to the first question (Wiggins's descriptivism might stand as independent justification for the physiological-functional account), but it is to this second question and the scientific clash with pre-theory (specifically in the form of biological reductionism) that attention will now be turned.

# Notes

1 Wiggins 1967: vii, 1980: vii, 1996: 228, 2001: xi.
2 Wiggins 2001: xiii, 2005: 475–6, 2012: 14ff.
3 Wiggins 2012: 14. Again, there are variant readings: contrast this with Jack Wilson's thought that 'Wiggins explicitly advocates a scientific approach to determining the substantial kinds' (1999: 34).
4 Wiggins 2001: 234 (see also 2005: 475–476).
5 Michael Ruse 1987: fn. 358.
6 E.g. van Inwagen 1990: 84.
7 See Wiggins's Leibnizian thoughts on this (2012: 8).
8 Wiggins 2001: 86.
9 E.g. Okasha 2002.
10 Dupré 2010/2012: 125. See also Paracer and Ahmadjian 2000 for a good overview.
11 Dupré (2008/2012):

> Our symbiotic microbes are essential to our well-being.... In fact, for the majority of mammalian organism systems that interact with the external world – the integumentary (roughly speaking, the skin), respiratory, excretory, reproductive, immune, endocrine, and circulatory systems, there is strong evidence for the coevolution of microbial consortia in varying levels of functional association. At any rate ... there is much to be said for thinking of the whole community that travels around with me as a single composite entity.
>
> (86–87)

12 Wilson 1999: 6–7.
13 Wilson 2000: s304.
14 Wilson 1999: 129.
15 E.g. the 'grex' discussed by Wilson in 1999 (8).
16 Haldane 1927: 286.
17 Jack Wilson (1999) draws attention to the tendency of such philosophers to overlook the relevance of philosophy of biology to personal identity – and it is a shame that his thoughts have had so little effect in the latter sphere (save for a small review of his book by Olson (2001)).
18 See, for example, Clarke *forthcoming*, Pradeu 2010, Wilson 1999.
19 Sober 2000: 151.
20 Pradeu, 2012: 230.
21 Gould 1984 (as quoted in Wilson 1999: 62).
22 Wilson 1999: 60f.
23 Wilson 2000: s302.
24 Pradeu 2010: 4.
25 Wilson 2000: s302.
26 Pradeu 2010: 5.
27 See Pradeu 2010, 2012.
28 Smith 1986, Dawkins 1976.
29 Where philosophers interested in personal identity *do* engage with this discussion this seems to be the line that they take – see, for example, Olson 1997: 129.
30 See Pradeu (2010: 5) for an elaboration of this picture.
31 Pradeu 2010: 6.
32 Dupré 2010/2012: 118.
33 Dupré 2010/2012: 118.
34 Harper 1977: 27.
35 Wilson 1999: 87.
36 Wilson 1999: 111.
37 Wilson 1999: 64.

38  See Dupré 2010/2012: 119ff for examples.
39  Giovanni Boniolo and Giuseppe Testa 2012, Boniolo 2013.
40  Okasha 2002.
41  Dupré 2012: 123.
42  Boniolo and Testa 2012: 279.
43  Boniolo and Testa 2012: 295.
44  As Boniolo (2013) puts it, an organism is a 'whole phenotype', which encapsulates

> all those phenotypic modules or units (the metabolic phenotype, the immunological phenotype, the nervous phenotype, the somatic phenotype, the behavioural phenotype and so on) that combine to make up what a human being is, at a given time-slice of his/her development.
>
> (2013: 403)

45  Boniolo and Testa 2012: 291.
46  Boniolo and Testa 2012: 295 (my emphasis).
47  Boniolo and Testa 2012: 283.
48  There are, note, further worries about whether the whole phenotype includes the extremities of what Dawkins calls the 'extended phenotype'. If the living being is its whole phenotype, understood as the result of the epigenetic processes that have regulated the expression of its genome, does this include – for example – the dirt mounds built by clonal insect colonies or the nests constructed by birds? One is also led to wonder whether 'superorganisms' exist on the epigenetic account: to what extent are coral reefs, for instance, living beings for Boniolo and Testa? It's unclear whether or not they have a scientifically grounded response to such questions.
49  Ruiz-Mirazo and Moreno 2012: 26. NB There is, I think, variance in the use of the term 'phenomenological' here (contrast, for example, the connotations and those expressed in Chapter 1).
50  Ruiz-Mirazo *et al.* 2000; Mossio and Moreno 2010.
51  Ruiz-Mirazo *et al.* 2000: 216.
52  Ruiz-Mirazo *et al.* 2000: 217.
53  Mirazo and Moreno 2012: 27.
54  Mirazo and Moreno 2012: 33.
55  Mirazo and Moreno 2012: 34.
56  Mirazo and Moreno 2012: 29.
57  It should be noted that the 'Basque groups' autonomy approach is but one of many. Another, which develops the position and moves beyond the 'metabolic' account described above is found in Matteo Mossio and Alvaro Moreno *Biological Autonomy: A Philosophical and Theoretical Enquiry* 2015.
58  Ruiz-Mirazo and Moreno 2012: 37. For Ruiz-Mirazo, to answer this one must compare *intra-cellular* (i.e. *within* the cell) and *inter-cellular* (i.e. *between* the cells) functional diversity and cohesion; the grex, like many bio-films, is a collection of cells that builds a common physical border and expresses basic motility – but otherwise, compared to the systemic complexity of the cells themselves, there is a limited degree of functional differentiation within the collective (2012: 37). The comparison leads one to privilege the cells 'autonomy' over the 'second-order' autonomy of the conglomerate.
59  Ruiz-Mirazo and Moreno 2012: 40.
60  See Pradeu 2010 and Pradeu and Carosella 2006. Though he does not, to my knowledge, endorse this position in his published works, Wiggins has done so to me in private correspondence.
61  Pradeu 2010: 2.
62  There are too many puzzle cases where it remains unclear whether the parts of the apparent organisms are themselves functionally integrated entities (Sober 2000: 151).
63  Pradeu 2010: 5.

64 Pradeu 2010: 4–5. It may be objected that only very few organisms exhibit immuno-genicity. This is an objection Pradeu thoroughly rejects: 'it is now clear to all immunologists that immunity is ubiquitous' (Pradeu 2010: 6).

65 Pradeu 2010: 9.

66 Paradigm texts include Metchnikoff 1907 and Loeb 1937 (more recently, Gould and Lloyd 1999).

67 Pradeu 2010: 9.

68 Pradeu 2010: 11.

69 One may wonder about ant's nests and other such items. On the phenomenal level these are seen to be collections of organisms, but might they be functionally integrated enough to be seen on the physiological-immunological level, as unities? If so, Wiggins will still be able to accommodate these entities without harming his everyday picture – as will be clear from the description of his verdant realism.

70 For an overview of the notion of developmental and historical individuals, see Wilson 1999: 60f.

71 Pradeu 2010: 4.

72 The genetic individual can be considerably smaller than the phenomenal one – and the two can separate like the object of diplopic vision (clones, for instance, cause the picture to blur). (It should be said that the genetic individual *is* sometimes taken to be the phenomenal one; consider, for instance, how DNA testing is used in legal cases – and problematically so. The occasional failings of this are troubling and well-documented (see, for example, Dupré 2010/2012: 118)).

73 Remember Wiggins's own suspicions about science-led metaphysics (mentioned at the start of this chapter and in Chapter 1); we should receptive but wary of scientific facts which appear to conflict with pre-theory.

74 This view of ourselves is decidedly local; as David Hull famously puts it, our commonsense view 'is strongly biased by our relative size, duration, and perceptual abilities' (Hull 1992).

75 Hull 1992: 184.

76 The controversy is, as he sees it, around which ontology is the most effective for certain purposes (e.g. discussions in molecular or evolutionary biology), not whether or not they exist) (Dupré 2012: 13).

77 Wilson 1999: 42.

78 Wilson 1999: 43.

79 Wilson 1999: 47.

80 As described by Wilson 1999: 46.

81 It should be mentioned that Dupré sees himself to be a 'pluralist' (see, for example, Dupré 2014) – but the view he endorses seems sufficiently similar to what I call open-minded realism for us to shelve his usage of the term. See Rasmus Winther for another persuasive account of 'interest-sensitive, pragmatic realism' (Winther 2011: 422–423).

82 Sortal essentialism is, roughly, the thesis that if an entity is actually individuated by a substance sortal in the actual world, it is necessarily individuated by that sortal. This is a feature of Wiggins's work, which is discussed extensively by Wilson (1999: 69f) and Penelope Mackie (1976). Mackie illustrates the thesis with an example about Aristotle: Aristotle falls under the substance sortal *human*. He cannot fall under this sortal and simultaneously fall under the substance sortal *centipede* – there are properties one must have if one is human which one cannot have if one is a centipede. Because Aristotle is a *human* and *human* and *centipede* are incompatible substance sortals, Aristotle could not have been a centipede. The sortals under which we fall determine, as Wiggins puts it, 'the counterfactual envisagings of states and histories alternative to *x*'s actual states and history' (2001: 123). See Jones 'From Individuation to Essentialism' (*draft*) for an extensive critique.

83 Okasha 2002.

84 Kripke talks of 'internal structure' and Okasha notes that there is an ambiguity here between genetic make-up and physiology. He denies the ambiguity will give Kripke leverage (Okasha 2002: 198). However, given the arguments above and below, one might think an anti-reductionist interpretation of 'internal structure' is still possible.

85 Okasha 2002: 207.

86 Okasha 2002: 196. While Okasha denies the existence of essential phenotypic traits, he claims that there *are* 'phenetic clusters', clusters of phenetic traits that tend to co-vary. One might wonder – as Michael Devitt does – whether this is just a more general form of essentialism (Devitt 2008: 371). Are the clusters not just the essences? Here the argument depends on how 'neat and tidy' one wants the essences to be – and I think there is something to be said for the essentialist's defence that the notion of a strict essence (or 'tiger gene') is a caricature of their position by their detractors ... but for the sake of the present discussion, I will assume the effectiveness of Okasha's position here.

87 Putnam 1975: 240.

88 Kripke 1980: 121.

89 Olson 1997: 124ff.

90 Wiggins 1980: 203.

91 Wiggins 2001: 143.

92 Note too that in *S&SR*, Wiggins explicitly distances himself from the sort of reductionism Okasha attributes to him and Putnam: 'It is', he writes, 'fully compatible with Putnam's suggestion ... that the theoretical description that comes into question in a given case should make reference to both the microphysical and the macrophysical' (Wiggins 2001: 80).

93 Is there such a thing as 'human nature'? Is there some real essence of the human species 'an internal property of all and only humans that explains why they are as they are and why they do as they do'? (Wiggins 2001: 1) It might be politically problematic to think so (on similar lines to those with which we critiqued the thesis of conceptual invariance in Chapter 1). These questions are the focus of Dupré's essay 'On Human Nature', in which he suggests that 'human nature', if there is such a thing, 'is the upshot of the interaction between a developing human individual and a particular society ... [it is] a matter of the constant interaction in human development between internal, narrowly biological factors, and external, generally social factors.' Wiggins, I think, would be amenable to this suggestion – and it appears, to some extent, in his comments about *Bildung*, described in Chapter 4.

# Bibliography

Boniolo, G. (2013) 'Is an Account of Identity Necessary for Bioethics? What Post-Genomic Biomedicine Can Teach Us', *Studies in History and Philosophy of Biological and Biomedical Sciences* 44, 401–411.

Boniolo, G. and Testa, G. (2012) 'The Identity of Living Beings, Epigenetics, and the Modesty of Philosophy', *Erkennetis* 76: 279–298.

Clarke, E. (*forthcoming*) 'The Multiple Realizability of Biological Individuals', in *Journal of Philosophy*.

Dawkins, R. (1976) *The Selfish Gene* (New York: Oxford University Press).

Devitt, M. (2008) 'Resurrecting Biological Essentialism', *Philosophy of Science* 75(3): 344–382.

Dupré, J. (2008/2012) 'The Constituents of Life 2: Organisms and Systems', in J. Dupré *Processes of Life* (Oxford: Oxford University Press).

Dupré, J. (2010/2012) 'The Polygenomic Organism', in J. Dupré *Processes of Life* (Oxford: Oxford University Press).

Dupré, J. (2012) *Processes of Life* (Oxford: Oxford University Press).

Dupré, J. (2014) 'A Process Ontology for Biology', on the *Auxiliary Hypotheses* blog, available online at http://thebjps.typepad.com/my-blog/2014/08/a-process-ontology-for-biology-john-dupré.html.

Gould, S.J. (1984) 'A Most Ingenious Paradox', *Natural History* (December): 20–28.

Gould, S.J. and Lloyd, E. (1999) 'Individuality and Adaptation across Levels of Selection: How Shall We Name and Generalize the Unit of Darwinianism?', *Proceedings of the National Academy of Science of the USA* 96(21): 11904–11909.

Haldane, J.B.S. (1927) *Possible Worlds and Other Papers* (London: Chatto and Windus).

Harper, J. (1977) *Population Biology of Plants* (New York: Academic Press).

Hull, D.L. (1992) 'Individual', in E. Fox Keller and E. Lloyd (eds) *Keywords in Evolutionary Biology* (Cambridge, MA: Harvard University Press): 180–187.

Jones, N. (*draft*) 'From Individuation to Essentialism'.

Kripke, S. (1980) *Naming and Necessity* (Oxford: Blackwell, revised edition).

Loeb, L. (1937) 'The Biological Basis of Individuality', *Science* 86(2218): 1–5.

Mackie, P. (1976) 'Sortal Concepts and Essential Properties', *Philosophical Quarterly* 44: 311–333.

Metchnikoff, E. (1907) *Immunity in Infective Diseases* (Cambridge: Cambridge University Press).

Mossio, M. and Moreno, A. (2010) 'Organisational Closure in Biological Organisms', *History of the Philosophy of the Life Sciences* 32: 269–288.

Mossio, M. and Moreno, A. (2015) *Biological Autonomy: A Philosophical and Theoretical Enquiry* (New York: Springer).

Okasha, S. (2002) 'Darwinian Metaphysics: Species and the Question of Essentialism', *Synthese* 131(2): 191–213.

Olson, E. (1997) *The Human Animal: Personal Identity Without Psychology* (Oxford: Oxford University Press).

Olson, E. (2001) 'Review of *Biological Individuality*', *The Philosophical Quarterly* 51(203): 264–266.

Paracer, S. and Ahmadjian, V. (2000) *Symbiosis: An Introduction to Biological Associations*, 2nd Edition (Oxford: Oxford University Press).

Pradeu, T. (2010) 'What is an Organism? An Immunological Answer', in P. Huneman and C.T. Wolfe (eds) *History and Philosophy of the Life Sciences*, special issue on *The Concept of Organism: Historical, Philosophical, Scientific Perspectives. History and Philosophy of the Life Sciences* 32(2–3).

Pradeu, T. (2012) *The Limits of the Self* (Oxford: Oxford University Press).

Pradeu, T. and Carosella, E. (2006) 'The Self Model and the Definition of Biological Identity in Immunology', *Biology and Philosophy* 21: 235–252.

Putnam, H. (1975) *Mind, Language, and Reality: Philosophical Papers* (Cambridge: Cambridge University Press).

Ruiz-Mirazo, K. and Moreno, A. (2012) 'Autonomy in Evolution: From Minimal to Complex Life', *Synthese* 185: 21–51. 26.

Ruiz-Mirazo, K., Etxeberria, A., Moreno, A. and Ibáñez, J. (2000) 'Organisms and their Place in Biology', *Theory Biosciences* 119: 209–233.

Ruse, M. (1987/1992) 'Biological Species: Natural Kinds, Individuals, or What?' *British Journal for the Philosophy of Science* 38: 225–42, repr. in Marc Ereshefsky (ed.) *The Units of Evolution: Essays on the Nature of Species* (Cambridge, MA: MIT Press), 343–361.

Smith, M. (1986) *The Problems of Biology* (Oxford: Oxford University Press, Oxford).

Sober, E. (2000) *Philosophy of Biology* (2nd Edition) (Boulder, CO: Westview Press).

van Inwagen, P. (1990) *Material Beings* (Ithaca: Cornell University Press).

Wiggins, D. (1967) *Identity and Spatio-Temporal Continuity* (Oxford: Blackwell).

Wiggins, D. (1980) *Sameness and Substance* (Cambridge, MA: Harvard University Press).

Wiggins, D. (1996) 'Replies', in S. Lovibond and S. Williams (eds), *Essays for David Wiggins: Identity, Truth and Value* (Oxford: Blackwell Publishing).

Wiggins, D. (2001) *Sameness and Substance Renewed* (Cambridge: Cambridge University Press).

Wiggins, D. (2005) 'Reply to Bakhurst', *Philosophy and Phenomenological Research* 17(2): 442–448.

Wiggins, D. (2012) 'Identity, Individuation and Substance', *European Journal of Philosophy* 20(1): 1–25.

Wilson, J. (1999) *Biological Individuality* (Cambridge: Cambridge University Press).

Wilson, J. (2000) 'Ontological Butchery: Organism Concepts and Biological Generalizations', *Philosophy of Science* 67, supplement, s301–s311.

Winther, R.G. (2011) 'Part-whole science', *Synthese* 178: 397–427.

# 7    Reduction and emergence

It is hard to resist the thought that an analysis of our metaphysical character should acknowledge the minuscule things which appear to compose us. Yet in attending to these – cells, bacteria, atoms, quarks and so on – we may find we are being drawn to think of ourselves in a way that conflict with our everyday experience. The most obvious and most dramatic reaction to these entities is *biological reductionism*. Impressed by these minuscule beings, and finding them to offer generative explanations of higher-level phenomena, scientifically suscepti-ble metaphysicians are apt to claim that such things are, ontologically speaking, more 'fundamental' than the organisms that they compose. In their zeal, these metaphysicians surrender to the passion for *ranking*. Entities at higher levels (organic substances, for example) are downgraded and seen as a sort of onto-logical run-off. In its coarsest form, this view construes us as 'epiphenomenal' beings, little more than atoms arranged 'human-being-wise'. Even in its weaker forms, the position contrasts markedly with our pre-theoretical, phenomenal par-titioning of the world. We do not experience ourselves – you or me – as mere bundles of particles. Individual particles are beyond our sense-range.

How might Wiggins construct his defence of the everyday ontology? We should note from the start that not all metaphysicians who interest themselves in microscopic entities endorse reductionism. There are some, John Dupré for instance, who argue *against* metaphysical reductionism and speak for a kind of *emergentism* which sees the smaller entities and the larger entities of our everyday awareness as having 'equal claims to reality'.[1] Dupré's *promiscuous realism* rejects the claim that organisms are metaphysically inferior.[2] Wiggins is a natural recruit to this position.

In the end we shall find this kind of emergentism to be at odds with the pre-theoretical picture, and thus with Wiggins's approach. While in Dupré's earlier work there are no claims about the metaphysical character of these various entities (they are all just equally metaphysically 'robust'), more recently he has advanced a form of *process* philosophy, akin to that of A.N. Whitehead.[3] It is no surprise that Dupré should endorse this view given how, time and again, biological investigations direct our attention to the *transience* of the matter that makes us up, and the sym-biotic activity that appears to constitute us. It is at this point that the descriptivist will take umbrage. Emergentism is one thing, Wiggins will say, and the processual view is another. For however much we depend upon processes, we are not processes. We are continuants that persist through space and time. The *processual* view that emerges from this form of emergentism is inimical to the pre-theoretical understanding of everyday beings that underpins Wiggins's metaphysics.

## 1  The reduction of the organism

'Reductionism' refers to not one, but a family of interrelated theses,[4] and like all such 'isms' it is disconcertingly vague, with nigh on as many definitions of the term as tokens of the word. For present purposes, we could do worse than let ourselves be guided by Keith R. Benson's somewhat neglected paper 'Biology's "Phoenix": Historical Perspectives on the Importance of the Organism'. Benson explains the twentieth-century relegation of the concept of the 'organism' to the backwaters of biological theory, and sets out an opposition between older, organismic theories of biology and more recent reductionistic ones, writing:

> [T]he older tradition, with its descriptive methods from Aristotle, and the modern interpretation, often described as reductionistic and mechanistic, have been depicted as mortal adversaries and competitors. J.H. Woodger, the major twentieth-century advocate of the organism, popularized this notion in his 1930 article in the *Quarterly Review of Biology*.
>
>> In histories of biology in the dim future there will probably be a chapter entitled 'The Struggle for Existence of the Concept of Organicism in the Early Twentieth Century', which will relate how this concept came to be neglected on account of the influence of Descartes, how the metaphysics of natural science in the Nineteenth Century so completely dazzled biologists that they never dreamed of regarding organisms as being anything but swarms of little invisible hard lumps in motion, and how the first blossoming of the concept of organism towards the end of the century was nipped in the bud by the mismanagement of those who advocated it.
>> (Woodger, 1930–1931)
>
> The struggle, which David Hull described as 'more reminiscent of political polemics and biblical exegesis than science' (Hull, 1974), eventually and inexorably led to the removal of the organism from centre stage by the twentieth century.[5]

This passage is interesting for a variety of reasons. It captures some of the passion, and the pseudo-moral tenor, of the debates about reductionism. It maps perfectly onto the analysis offered above. Wiggins takes organisms to be paradigm substances; he falls firmly on the 'Aristotelian' side of the described opposition. He exemplifies the 'descriptive methods' of Aristotle, where metaphysics is guided by our everyday navigation of the world, and focuses on entities such as human beings to which we cannot avoid attending. Wiggins is an anti-reductionist in precisely Woodger's fashion, denying that we are simply 'swarms of invisible hard lumps'. The mention of J.H. Woodger, a theoretical biologist and philosopher of biology, is also significant; Woodger is one of the few to whom Wiggins regularly refers.[6] In the anti-reductionist cause it is Woodger's Tarskian explication of reduction that Wiggins reproduces in *S&S* and *S&SR*.[7]

Benson's article, written in 1989, situates reductionism as the mainstream. The terms 'reductionism' and 'anti-reductionism' are clarified more precisely below, but Woodger's characterization captures the general idea: reductionists see organisms to be nothing more than the sum of their physico-chemical parts; in the end, human beings like you or me are somehow less real than the tiny entities, the mereological simples, that make us up. Wiggins is not alone in resisting this picture. Towards the end of his paper, Benson suggests that imminent research will encourage a revival of interest in anti-reductionism, and this has been borne out, to a significant extent, by the now numerous critiques of reductionism and the growing popularity of the kind of emergentist thesis found in the work of John Dupré.

## 2 'Reductionism' and 'emergence'

Ingo Bridgandt and Alan Love present a helpful tripartite division between types of reductionism: *epistemic*, *metaphysical* and *methodological*.[8] As they construe it, *epistemic* reductionism focuses on theories, or concepts or models, and claims that such items in one scientific domain – e.g. biology – can be *translated into*,[9] or *derived from*,[10] or *explained by*[11] theories, concepts or models in another, say physics. Here, reduction can be conceived *synchronically* or *diachronically* – between contemporaneous theories or successive ones.[12] Epistemic reductionists are primarily interested in how the branches of human knowledge connect, *not* in what entities there are or what methods we should employ to investigate the biological realm.

*Metaphysical* reductionism focuses on the structure of reality and (relatedly) whether certain biological items do or do not exist. In general terms, the metaphysical reductionist's thesis is that biological things are 'nothing more' than the physico-chemical items that make them up (bundles of quarks or fields, etc.).[13] The historical debate about the viability of *vitalism* demonstrates these kinds of metaphysical concerns. The vitalist holds that there are non-physico-chemical forces – vital sparks or *élan vital* – that govern biological systems.[14] The metaphysical reductionists (in this context sometimes called 'mechanists')[15] reject this view of non-spatial, formative 'entelechies'. Nowadays, most philosophers of science do the same. Few endorse mysterious *vis vitalis*.[16]

*Methodological* reductionism relates to scientific practice. The methodological reductionist claims that biological research should work 'from the bottom up'. One should look to the lowest, most fundamental levels to understand the higher-level features. Biological systems, including organisms, are best investigated by looking for molecular and biochemical causes. This kind of strategy is described by Bechtel and Richardson as 'decomposition and localization',[17] and is often articulated by reference to the use of the *machine* motif, ubiquitous since the seventeenth century, as the 'founding metaphor of modern science'.[18] The methodological mechanist (described by Dupré, Grene and Depew, Toulmin and Goodfield),[19] states that items in the biological realm should be understood in the same way as we understand human-engineered machines – that is by looking to the parts and working upwards. (This, as we will see in Chapter 9, feeds into the notion of the *biological artefact*.)

However gestural, this tripartite analysis provides a general framework for separating out, and assessing, the various reductionist claims.[20] In his paper, Benson suggests that reductionism is in a state of decline. As regards *epistemic* and *methodological* types, this predicts well enough the current consensus among philosophers of biology.[21] Over the past decades, there has been growing doubt about epistemic, 'theory' reductionism, doubt motivated by issues with *multiple realizability*[22] and the *deductive-nomological model*.[23] Similarly, despite the successes of a mechanistic methodology, it is now seen to be highly controversial, systematically blinding the enquirer to relevant high-level features in biological systems.[24]

*Metaphysical* reductionism, on the other hand, is still the subject of lively debate – and this is our concern here. It is the metaphysical reductionists who are so enamoured of the microscopic beings that they have to assign to human beings some kind of secondary metaphysical status. 'The reductionist', as Dupré has it, 'believes that in the end there is nothing in the world but the stuff of which things are made … this basic physical stuff.'

> Of course, the reductionist does not say, bluntly and absurdly, that houses, for example, don't exist. The claim is rather that a house is, ultimately, nothing but an aggregate of physical stuff, and all the properties of any house can, in principle, be fully explained by appeal to the properties and relations of basic physical stuff. So there is a possible, microphysically grounded, account of the world which would have no need to mention houses.[25]

As Dupré presents it, the reductionist sees organisms (and houses and so forth) as nothing more than the physical entities (quarks, etc.) that make them up. For the reductionist they do not 'really' exist in the way that the basic physical stuff does. Elsewhere, Dupré puts this in terms of 'ontological priority'[26] and 'ontological primacy'.[27] The reductionist views homogenous matter as more 'metaphysically robust' than higher-level objects. This is an unnerving claim, given the apparent solidity and dependability of chairs and trestle-tables and trees.

While many have abandoned the 'epistemological dream of reductionism',[28] discussion still continues about metaphysical reductionism and the reality of biological items.[29] The initial appeal of the thesis is clear. It aligns itself with the materialist position that the only things that exist are material things), a view that is undeniably, and understandably, popular. There are no vital sparks, or entelechies, or other such mystical forces. Moreover, there is something markedly attractive in the idea that – even if it lies beyond our conceptual reach – a full and exhaustive description of reality might *theoretically* be given in terms of these fundamental building blocks. Reductionism presents a pleasingly ordered picture of reality. Of course, it may fail to marry up to our pre-theoretical framework. The child finding her way in the world need hardly attend to the mereological simples, let alone avow their existence. But so much the worse for the child, the reductionist may say. She will learn the truth in time.

Over and above talk of ontological priority and primacy, philosophers often construe the debate in terms of 'reality'. The reductionist holds that physical stuff is *real*, while questioning the reality of higher-level items. And there is a special notion of 'reality' in play here. *To be real*, they say, *is to have causal powers*. This is the view formalized in Jaegwon Kim's influential 'Alexander's dictum'.[30] It sits as the pivot around which spin many of the current, metaphysically accented discussions of reductionism. Reductionists hold that biological entities are not real in the way in which the basic physical stuff is. The properties of those higher-level systems are, in the end, simply the result of a linear causal chain reaching up from the material substrate. It is not denied that organisms have causal powers but these are in a sense derivative. They are derivative in the same way as the power of a shoe to kick a ball derives from the power of the foot that moves within it.

The picture is neat and tidy. Still, there are those – like the descriptivists, with their respect for pre-theory – who try to resist it, and it is this claim about causality that constitutes the main point of conflict between philosophers of biology. John Dupré is prominent among defenders of the anti-reductionist view. In *Processes of Life*, he says 'I am insisting ... that there is a whole hierarchy of increasingly complex things that really exist, and that have causal powers, that are not reducible to the mechanical combination of the powers of their constituents.'[31]

> At the basis of [my] position is the idea that many or all such entities have causal powers that are not simply consequences of the way their physical components are fitted together.[32]

Dupré claims that, when physical systems reach certain levels of complexity, *novel* causal properties emerge, not deducible or derivable from the properties of their parts.[33] Organisms (among other things) have these novel causal properties and are just as real as their constituents. His *emergentist*[34] arguments invoke a considerable amount of biological research into, for example, processes like *methylation*. This process (already described in Chapter 6 in relation to the *epigenetic* account) is the means by which higher-level factors such as changes in the cellular environment alter the behaviour at the lower-levels – e.g. the conduct of DNA molecules. Here, we see an inversion of the reductionist's causal story: no reality without causality but there are causal powers that exist at the higher levels. For a classic example consider the maternal care of infants in rodent populations.[35] These have been shown to produce methylation of genes in the hippocampus. A more prosaic example of this 'downward flow of causal influence'[36] is found in Dupré's description of the causal powers of a street market, like the

Albert Cuypstraat. Of course, the market depends for its existence on the people who go there. But it is more than a collection of people; it is a social institution with its own rules and regulations. It is part of the power of the market that it attracts the people who constitute it. Moreover, it determines what these people eat for dinner, through the produce that is sold there.

It is not my purpose to rehearse all the arguments and counter-arguments for reductionism here. The point is rather to capture, firstly, the balance of the dialectic between the reductionists and the emergentists, and then the timbre of Dupré's response. Doing so, we can see where Wiggins might figure in this debate. The reductionist's view is inimical to his common-sense approach. Might he then side with the emergentist?

## 3  An emergentist reading?

There are certainly elements of his project that suggest he might be sympathetic to some version of the emergentist view. Notable in this regard are his references to the biologist J.Z. Young in both Prefaces to *SS* and *S&SR*. In the latter he writes:

> I shall recall from the 1980 Preface the keen pleasure that I felt at that time on discovering how, in response to all the facts that confront the biological scientist, Professor J.Z. Young had arrived, in chapters Five and Six of his *Introduction to the Study of Man* (Oxford, 1971), at a conception of identity and persistence through time that is strikingly similar, where living things are concerned, to the neo-Aristotelian conception that I defend:
>
> > The essence of a living thing is that it consists of atoms of the ordinary chemical elements we have listed, caught up into the living system and made part of it for a while. The living activity takes them up and organizes them in its characteristic way. The life of a man consists essentially in the activity he imposes upon that stuff ... it is only by virtue of this activity that the shape and organization of the whole is maintained.[37]

'The life of a man consists essentially in the activity he imposes upon that stuff.' Young's choice of the word 'imposes' suggests a causal influence that does not derive from the body's constituents. In another passage that Wiggins quotes, Young writes:[38]

> Thus the activity which we hold to be characteristic is not expressed by describing DNA alone, since the interchanges between the organism and the environment are the factors that determine what sections of DNA are to be transcribed at a particular time and the rates of transcription.[39]

Here, Young alludes to a key challenge that is made against the reductionist program, a problem characterized by David Hull, as the 'one–many' objection.[40]

As Young indicates, the transcription of the DNA depends on the *context* in which it occurs. This view is now widely held.[41] The implication is that the functions of the molecules can only be understood in terms of *context*. And the context is the *organized whole* in which they occur.

This last point counts as a criticism of practical 'epistemic' reductionism, but the passage can also be read as an endorsement of metaphysical anti-reductionism. This is to say that Young can be interpreted as claiming that the activity of the whole exhibits properties that are not exclusively derived from properties at the genetic level. The activity of the organism is not causally speaking downstream from the activity of its parts. This reading can be supported by Young's own analysis of homeostasis and 'living activity' in Chapter 6 of his *Introduction to the Study of Man*.[42] It is well captured by his comments at the start of Chapter 3 ('Living Organization'). He states:

> Study of the outlines the elementary molecular composition of living things ... has already shown some of the characteristic features of organisms. But analysis only at that level can never provide the full basis that we require for forecasting their behaviour. In every known organism the molecules are organized into systems of a higher order of complexity, the cells.[43]

Even though Young himself does not detail explicitly his anti-reductionist position, these passages, alongside his criticisms of epistemic reductionism,[44] amount to something very close to biological emergentism. It is one of the several sources of Wiggins's general inclination towards emergentist thought. This, at least, is one line of interpretation.[45]

Wiggins's emergentist sympathies are evident elsewhere – in his emphasis on *Bildung*[46] and his claim that languages (and other social items) are irreducible public objects. These thoughts, which appear in his reply to David Bakhurst (already referred to in Chapter 4), suggest that he finds novel causal properties at higher levels than the physico-chemical. On Wiggins's view – as on Bakhurst's – human beings have distinctive psychological capacities, including rationality and morality, which are the effect of *cultural formation*. (This emphasis on our 'encultured' nature recurs frequently in his texts.[47]) He claims that our conception of ourselves should embrace 'our capacities to assimilate culture [and] our achievement, such as it still is, of Bildung'.[48] In saying this, he explicitly positions himself in opposition to what he calls the 'reductive fixations of ... present day biological science'.[49] He takes the process of enculturation to stand against the kind of genetic determinism that reduces morality to a product of genetic evolution.

In the reductive picture – as found, for example, in the work of E.O. Wilson[50] – our psychological capacities (and their causal powers) are the result of an upwards causal flow from the interactions of the more fundamental parts.[51] Wiggins, in contrast (under the influence of Bakhurst[52] (himself under the influence of the Russian psychologist, Lev Vygotsky[53])) specifically ties the development of our psychological traits to our assimilation of culture.[54] That is to say (as Bakhurst puts it), that it is 'through participation in and internalization of social forms of activity that [a] child's mind is *created*'.[55] This thought surfaces again and again in Wiggins's essays on human value (in his collection *Needs, Values, Truth*).[56] Our psychological capacities are not simply the mechanical run-off from some genetic blueprint. Rather, he contends, they depend on and feed into 'a process of interpersonal education, instruction, and mutual enlighten-ment'.[57] There are causal influences other than our genetic make-up which signifi-cantly shape the development of a human mind (if not its genesis)[58] – specifically, the systems into which it is initiated via language and culture.[59]

Two points can be made here. First, these comments by Wiggins – in par-ticular, his reservations concerning 'present day biological science' – account for his wariness (stated at the start of the previous chapter) about the wisdom of turning to biology to articulate the *human being principle*. Where biology is taken to examine only the effects of the lower-level causes, its focus is too narrow.[60] Second, and more importantly, Wiggins's emphasis on the formative effects of culture again suggests sympathy for the emergentist's thesis. The con-nection is strengthened further when it is remembered that Wiggins endorses a libertarian account of free will.[61] His suspicions of Dawkinsian 'reductive fixa-tions' is all of a piece with his hostility to determinism. Worries of this sort are typical of emergentist positions.[62]

Yet, in itself, the focus on *Bildung* is not emergentist. It leaves open the theoret-ical possibility that these cultural systems, and everything else, can ultimately be understood in purely physico-chemical terms. It is important then that Wiggins appears to deny that these over-arching systems of language and 'culture' (where culture is 'education in the broad sense')[63] *are* reducible to the psychology of the individuals from which they arise. The claim about the irreducibility of language may be found in his comments in 'Language as a Social Object':

> [A] language like English or Polish is a social object, a public thing with attributes irreducible to the individual psychology of its speakers.[64]

And the irreducibility of culture – in this instance, of the human susceptibility to, and penchant for, humour – is found in 'A Sensible Subjectivism':

> What is improbable in the extreme is that, either singly or even in concert, further explanations will ever add up to a *reduction* of the funny or serve to characterize it in purely natural terms (terms that pull their weight in our theoretical-cum-explanatory account of the mechanisms of the natural world). If so, the predicate 'funny' is an irreducibly subjective predicate.[65]

(Note that in this latter passage, Wiggins seems to be saying that it is, *in prin-ciple*, irreducible – not that its irreducibility is a contingent fact of our limited knowledge.)

Swift though this analysis is, we can see how the combination (of the causal power Wiggins attributes to language and culture and their apparent theoretical irreducibility) invites an emergentist reading. Seen alongside the endorsement of Young, and Wiggins's own libertarian commitments, this interpretation is strengthened further. Can it be that, in resisting the excesses of biological reductionism, Wiggins will position himself exactly alongside Dupré?

## 4 Reasons for doubt

The brief answer, I think, is 'no'. Despite growing popularity,[66] emergentism remains deeply controversial and Wiggins will be wary of endorsing a thesis that raises worries about causal exclusion and downwards causation.[67] It is sufficient here to note that, while critical of epistemic reduction,[68] Wiggins also, at points, appears to doubt the occurrence of novel causal properties at higher levels. Consider his apparent endorsement of *supervenience* – the thesis that any change at a higher level corresponds to a change at the lower level (the higher *supervenes* on the lower).[69] In *S&SR* he writes:

> [T]he kind-bound laws of coming to be, of distinctive activity, and of passing away are nomologically grounded. They are *supervenient upon*, or better (as Leibniz might put it) *consentient with*, the more basic laws that are immanent in all things.[70]

And again, in 'Identity, Individuation and Substance', he talks of the organismic as

> a (level or category) of being that the ultimate constituents of reality subvene/sustain/make possible.[71]

While supervenience is sometimes conceived of as non-reductive,[72] Dupré points out that it typically corresponds to a form of *practical*, but not *theoretical*, non-reductionism.[73] For the supervenience theorist, while we humans may fail to grasp the link between the lower levels and the higher, it is in principle possible for this link to be understood (by, for example, a divine mind). Wiggins's apparent endorsement of supervenience – when seen alongside Dupré's reading of that position – may suggest he is both aware of the criticisms of emergentism and, to some extent, persuaded by them.[74]

The emergentist reading also provokes certain methodological problems. Dupré, for example, construes the emergentist/reductionist controversy in terms of the ontological 'primacy' or 'priority' of some items over others. At times he also talks of claims to 'reality'. Harking back to the distinction drawn between the Aristotelian and Quinean projects, one may wonder from which of these alternatives his own emergentist's thesis issues. 'Primacy' and 'priority' imply some notion of ontological *ordering*, a hierarchical approach typical of the Aristotelian view. At the same time, there is a distinctly Quinean tone to Dupré's use of 'reality'. It is easy to read questions about what is and what is not *real* as questions about what does and does not *exist*.[75] In *The Disorder of Things*, at least, emergentism seems to attach to a Quinean conception of a flat ontology:

since organisms have novel causal properties they should be included on the ontological call-sheet.[76,77]

It is an obvious point, but one worth repeating, that in order to substantiate their position emergentists like Dupré also appeal to science. Science will tell us where the causal buck stops[78] and is positioned as an arbiter of ontological questions. It is to science that the emergentists turn when settling metaphysical disputes.[79] In this respect the emergentist position corresponds to the *revisionary* approach described in Chapter 1. Emergentists stand to be wrong about whether these ordinary, everyday objects are real. Engaging with the reductionists, emergentists run the risk of involving themselves in ontological revisions. This Quinean-cum-revisionary framework, which we find in Dupré's earlier work, contrasts noticeably with Wiggins's neo-Aristotelian, descriptive account. Wiggins is not interested in putting ticks or crosses in an existence column – rather, his aim is to examine the metaphysical character of the *substances* we find around us. His project is guided, not so much by science, as by our everyday interaction with the world.

This leads us to a more interesting concern. Attending to Dupré's recent work – including the essays found in *The Processes of Life* – one encounters a shift in focus. He and his set have begun to examine the metaphysical character of biological individuals. Among these philosophers there is still some resistance to the graded Aristotelian view. There is, of course, the same deference to science, but there is a new interest in the ontological profile of the entities under investigation. In acknowledging the constant movements, collaborations and fluctuations of these various tiny things, Dupré and others have started to endorse a *processual* view of biological individuals. For science, with its microscopic gaze, sees not static substances, but temporary swirls and eddies 'in a flux of change'.[80]

> It is arguable that our thinking is for deep reasons anchored to conceptions of objects describable in static terms; certainly many biological concepts are described in such static terms. But such concepts can only capture particular frozen time slices through the more fundamental processual biological reality.[81]

> The organism is a process.[82]

In the first passage, Dupré recognizes how a view of organisms as processual beings conflicts with our everyday understanding of the world; the opening sentence can be read as a claim about pre-theory and its apparent limits. But then he advances another position. Given the insights of the sciences into organic systems he offers a refinement of our ordinary understanding of ourselves and other animals: *the organism is a process*. The move is revisionary and consciously so.

It is not inevitable that the emergentist position should ally itself with this *process* view. Nonetheless, this is the tone of the discussion as it currently stands. Here then is another obstacle to Wiggins's endorsement of Dupré's form of anti-reductionism. He will surely say: *we do not pick ourselves out as processes*; nor do we see ourselves as non-discrete pulsing nodes in causal nexi. We think of each other as stable, solid beings, with determinate heights and weights and so on. We think of ourselves as three-dimensional beings, that is, as *continuants* who travel *through* space and time. We do not think we are spread out *over* time, as processes seem to be,[83] with different temporal parts in different temporal locations. So, I think, Wiggins will say.

Much here rides on what exactly is meant by 'processes'.[84] It bears noting that there is vigorous debate at the moment about the possible confusion of continuants and what Rowland Stout calls 'occurrents', which include processes.[85] (This debate, furthermore, is largely unremarked within the sphere of the philosophy of biology.) An occurrent is a being that *occurs*, or *happens*, of which the paradigm is an 'event', like a football match. An occurent is extended *through* time. It has temporal parts (its kick-off, for instance) in the same way that we have spatial parts (limbs in different spatial locations). A football match – unlike a continuant – does not progress through time *in its entirety*. Rather it has different parts in different times; its start, the kick-off, stands at 4 p.m., and its end, the penalty shoot-out, occurs at 5.30 p.m. As Helen Steward makes clear, in her beautifully precise paper 'What Is a Continuant?',[86] things such as football matches *do not change*. They cannot, because they do not pass through time; their temporal parts are fixed.

> The event itself does not change, any more than an apple changes which is redder on one side than on the other. It is merely that some of its parts – in this case, temporal, as opposed to spatial parts – possess properties which are different from those of certain other of its parts.[87]

There is some disagreement about whether processes are distinct from events as a form of occurrent, but the thought, which Steward emphasizes in her paper, is that processes (like the living process) are still understood as having *temporal parts*. Processes should not then be confused with continuants (such as particular human beings). Consider the process by which some zygote developed into an adult human. The process itself did not continue *in its entirety* through time. There was the 'germination' part of the process, restricted to a specific time (very near the start), then there was the 'puberty' part and so on. These parts of the process are not all present at once, in the way that my spatial parts are. It may be, as Stout argues, that processes, unlike events, can be said to *change* through time. The germination might be fast at t1 and slower at t2; in this way the germinating has a property at t1 it loses by t2. But as Steward indicates, susceptibility to change is not a sufficient condition for being a continuant. I am overlooking subtleties here, but the pivotal fact, which will prompt Wiggins to distance himself from Dupré, is that processes have temporal parts and despite the best efforts of David Lewis and other four dimensional theorists, we simply do not and cannot see ourselves in that way.

It is gratifying to find that these concerns appear, if somewhat obliquely, in a paper of Wiggins's, in the 1982 collection *Language and Logos*: 'Heraclitus' conceptions of flux, fire and material persistence', describes precisely this worry about the confusion of continuants and processes:

> Heraclitus writes happily of 'rivers', 'souls', 'the barley drink' – of continuants, that is. To insist that he really thinks of these things as processes, not as continuants, is to try to make a contrast that is quite anachronistic – and, on top of that, a category mistake. Processes are regular or gradual or fitful, take time, have temporal parts.[88]

And again, we find the thought expressed, in his 2012 paper, 'Identity, Individuation and Substance' (his target is the four-dimensionalist, but the critique can be extended to the process view – and indeed, perhaps calls us to question the level to which those positions are genuinely distinguishable):

> A conscious being cannot think of itself – or of the persons, organisms or physical objects that it encounters – as having the shape of a complete succession of events and present only in part at the moment of action or deliberation. Nor can it think of itself as an event or an event in the making. To act and think as it does, a conscious being must think of its whole self as present at the moment of reflection, perception or action and poised to persist in that way in the future.[89]

We surely do pick out processes in our everyday lives – ageing and falling and growing and so on – but we do not pick *ourselves* out as processes. This, at least, is Wiggins's view. There are continuants and there are processes, and to confuse the two as certain readers of Heraclitus have done, as Whitehead did, and as Dupré perhaps is doing, is to make a *category mistake*. If we conceive organisms as processual beings, we misunderstand their metaphysical character. We are not beings with temporal parts; we move through space and time in our entirety. (We may lose parts, but this counts as a determinate *loss* – we do not *lose* temporal parts.) This is not to say that processual beings do not exist, but that they are far from the main concern of Wiggins's project. We find this emphasized, again, in *S&SR*:

> The answers that these questions [of individuation] require from philosophy ought to be given in a language that speaks as simply and directly as natural languages speak of proper three-dimensional continuants – things with spatial parts and no temporal parts, which are conceptualized in our experience as occupying space but not time, and as persisting whole through time.[90]

*   *   *

Metaphysicians who turn to biology are treated to a kaleidoscopic picture of organic life. They find a host of tiny particles performing intricate and spectacular dances, and they respond to this display in a variety of ways. Some say that the organism is less fundamental than the dancers that make it up; others abstain from metaphysical pronouncements; others still claim that it is the dancing of the particles that is real – we are processual beings, they say. Each of these responses jars knowingly with any version of Wiggins's descriptivist project. Behind each lies the thought that science should legislate revisions.

How will Wiggins respond to the metaphysical reductionist? While initially inviting, emergentism has come to seem problematic. Its new association with processualism is rebarbative to the pre-theoretical understanding. In the next chapter I suggest that Wiggins can avoid this turmoil by reinforcing his open-minded metaphysical realism and developing his concept of *substance*. There are different, cotenable, metaphysical projects; there is the science-led one, which (as evidenced by the emergentist/reductionist discussion) focuses on *causal* dependence. But there is another one, an older one, an *Aristotelian* project, which focuses on our everyday pre-theory, and on *ontological dependence*, and may be no less sustainable.

## Notes

1 Dupré 2008a/2012: 70.
2 'Cats and dogs, mountains and molehills … exist in just as metaphysically robust a sense as do electrons and quarks' (Dupré 1993: 89).
3 For a recent, brief statement of Dupré's processualism, consider his entry on the Auxiliary Hypotheses blog: http://thebjps.typepad.com/my-blog/2014/08/a-process-ontology-for-biology-john-dupré.html.
4 Dupré 1993: 88.
5 Benson 1989: 1067–1068.
6 There are references to his *Biology and Language* (Woodger 1952) and his *Axiomatic Method in Biology* (Woodger 1937) in e.g. Wiggins 2001: 156 and 38. Until recently, Woodger has been rather neglected in the philosophy of biology literature – but this is a neglect now being rectified (see Nicholson and Gawne 2013).
7 Wiggins 1980: 48, 2001: 156.
8 Brigandt and Love 2008. See also Nagel 1998, Sarkar 1992 and Wolfe 2010.
9 *Translative* reductionism is the kind of reductionism found in the work of the positivists, like Otto Neurath and Rudolf Carnap. In the *Unity of Science*, Carnap advocates the translation of all statements – phenomenalist, biological, philosophical, etc. – into the language of the physical sciences (Carnap 1934). The aim was the creation of a common language in which truly interdisciplinary discussions could be conducted and redundancy between theories eradicated (this was for practical, as well as ideological reasons – Cartwright *et al.* 1995).
10 The decisive move away from Carnap and Neurath's positivist reductionism occurs in Ernest Nagel's *The Structure of Science* (1961). Nagel's model – *derivative* reductionism – holds that a reduction is effected when the laws of the target theory are shown to be logical consequences of the theoretical assumptions of the base theory (Nagel 1961: 345–358).
11 *Explanatory* reductionism – advanced by, for example, Wimsatt and Kenneth Waters (Wimsatt 1976, Waters 1990) – states that the target domain reduces to the base domain when the latter explains all of the observations that are explained in the former. Like the derivative reductionists, the explanatory reductionists posit a single 'ultimate' theory but also (i) admit reduction between fragments of theories *and* individual facts (thus taking theoretical weight off problematic 'bridge laws') and (ii) describe reductive explanations as *causal* explanations, where higher-level features in the domain of, for example, biology, are explained by the interaction of the *constituent parts* in the base theory (see, for example, Kauffman 1971) (thus avoiding

involvement with the controversial deductive-nomological method (Theurer *forth-coming*: 3, Rosenberg 2003: 3)).

12  Dupré 1993: 94.
13  Brigandt and Love 2008.
14  The most well known representatives of this approach in the twentieth century are Henri Bergson and Hans Driesch (see Mayr 2004: 23), who describe the 'formative power', as ' "something" without spatial character and to which no definite position in space can be assigned', an entelechy (Driesch 1908).
15  Here, as with 'reductionism', we must recognize the ambiguous ambit of 'mechanism' and 'vitalism' (see Wolfe 2010 and Hein 1972) and resist being drawn into unhelpful dichotomies (e.g. vitalism *versus* mechanism (Roe 2003)). These thoughts, with respect to mechanism, will be clarified in Chapter 9.
16  Cf. Normandin and Wolfe 2013. The relevance of this historical position is also discussed in the Appendix.
17  Bechtel and Richardson 1993.
18  Dupré 1993: 2.
19  Dupré 1993, Grene and Depew 2004: 36, Toulmin and Goodfield 1962: 207–334.
20  It is, it should be noted, difficult for those of a descriptivist bent, to separate out the metaphysical and epistemic aspects – for the reasons described in Chapter 1 (metaphysical and epistemological issues are not as clearly distinguished when one thinks that metaphysical inquiry is guided by examination of the human mind).
21  Dupré 2010/2012: 129.
22  For example, in order for the psychological property *pain* to be reduced to physical properties all descriptions of being in pain, for any animal, will have to be worked out as descriptions of the possession of those physical properties. But it seems highly unlikely that there are such specific physical properties (see Putnam 1975: 436).
23  The deductive-nomological model of explanation proposes that phenomena should be explained by deduction from initial conditions and genuine, general 'laws of nature' – and there are now well-known counter-examples to this proposal, which show that this 'DN' model provides neither *necessary* nor *sufficient* conditions for scientific explanation (see Theurer *forthcoming*: 3, Rosenberg 2003: 3. See also, e.g. Scriven's 'singular causal explanations' which make no appeal to generalized laws (Scriven 1962) and Salmon's examples of 'defective explanation' (1989: 34)).
24  See, for example, Wimsatt 1980.
25  Dupré 2008a/2012: 72. For the full import of 'in principle' see Dupré's discussion of 'practical' versus 'theoretical' reductionism in Dupré 1993 (95–96). This passage also indicates certain points where the Brigandt and Love's tripartite division fails to identify salient connections between epistemic and metaphysical reductionist claims.
26  Dupré 1993: 89.
27  Dupré 1993: 92.
28  Dupré 2008a/2012: 70.
29  Wolfe 2010: 2.
30  Samuel Alexander was one of the pivotal figures in 'the Golden Age of British Emergentism' (see Malaterrre 2013: 160). For a more precise statement of Kim's dictum see Kim 1992 and 1993.
31  Dupré 2008a/2012: 72.
32  Dupré 2008a/2012: 70.
33  See Garrett 2013 for a discussion of how non-derivability relates to non-deducibility and inexplicability.
34  Note that 'emergentism' has numerous valences – see O'Connor and Wong, 2009 (cf. Mark Bedau 2003). See also Garrett (2013: 135) for further indication of the shortcomings of Brigandt and Love's epistemic/metaphysical division, particularly with respect to how epistemological claims of emergence relate to metaphysical ones.
35  Dupré 2012: 82–83.

36 Dupré 2012: 72.
37 Wiggins 2001: xi.
38 Wiggins 1980: 207. Note also, that this quotation appears alongside a quotation from another twentieth-century organicist – Joseph Needham (see the discussion of Needham in Haraway 1976).
39 Young 1971: 88.
40 See Hull 1972.
41 Indeed, it is often pointed out that the same allele may lead to two different phenotypes occurring in two individuals with a different overall genotype (Brigandt and Love 2008).
42 Young 1971: Chapter 6 – see particularly §1, §5, §6.
43 Young 1971: 37.
44 Boycott 1998.
45 Wiggins took Young in this way. But not everyone has. Olson and van Inwagen defer to Young, and yet both are – at points, explicitly – advocates of a form of biological reductionism. (It is interesting to note, in passing, the particular place that J.Z. Young inhabits in Anglophone philosophy. For some reason – perhaps related to his readability – he is one of the most frequently cited biologists in the personal identity literature. Young's metaphors and his account of biological processes are still in common currency. Curiously, he seems also to have stood as one of Daniel Dennett's PhD examiners.) Of course, they only implicitly endorse him (unlike Wiggins) but I think the point still stands.
46 Wiggins 2005: 475.
47 Wiggins 1987/1991: Essay V and 2001: Chapter 7 *passim*.
48 Wiggins 2005: 475.
49 Wiggins 2005: 475.
50 Wilson 1975.
51 Consider an (extreme) reductive account of our psychological capacities (taking morality and rationality as place-holders for the rest). A reductionist will describe these faculties and their properties as ultimately caused by, and exhaustively explicable in terms of, activity at the genetic level. We inherit a genetic 'program' (see Smith 2000, Griffiths 2001 and Avise 2001 for more on this metaphor) from our parents, which has been honed over the millennia by natural selection. This 'program' causes a particular kind of RNA production, which in turn creates certain proteins, which then organize the specific neurological structures that determine the psychological traits we enjoy. In the end, as Robert Wright has it, 'everything boils down to the genes' – including our psychological capacities (Wright 1994: 9). See also Dawkins, 2006 (217), Rose *et al.* 1984, Lewontin 1993 and Rosenberg 2006).
52 Wiggins 2001: 195, fn. 3.
53 See, for example, Bakhurst 1991.
54 See Lovibond 1996 for an overview.
55 Bakhurst 1991: 78.
56 It is the thought that underpins what Lovibond calls his 'Bildung model of value-experience' (1996: 76).
57 Wiggins 1987/1991: Essay V.
58 At points Wiggins denies that the human person is *created* through the process of socialization, enculturation, etc. Instead, he says that the human person is

> a creature with a natural capacity, which may or may not be realized, for reason, morality…. Adapting a dictum of Woodger's I have quoted before (see S&SR p. 64), would it not be better to say that the child is the primordium of the moral/rational being but not of the future person, because the child already is that person?
>
> (Wiggins 2005: 475)

168    *Reduction and emergence*

59  Bakhurst 2005: 467.
60  Thus, replying to Bakhurst, he asks 'how broadly ought the biological to be conceived' (Wiggins 2005: 476).
61  E.g. Wiggins 1987/1991: Essay VIII.
62  See, for example, Malaterre 2013: 158.
63  Lovibond 1996: 78.
64  Wiggins 1997: 499. See also Charles Taylor's discussion of Wiggins's anti-reductive view of culture in Taylor 2003.
65  Wiggins 1991: 195–6 (see also 197f., and 352f. for like-minded comments on sociobiology: 'the thing we really need to try to describe is what morality *has become*, a question on which evolutionary theory casts no particular light').
66  Malaterre 2013: 156–157.
67  Kim's 'exclusion argument against non-reductive physicalism states that since there is only one sufficient cause for any effect, if all physical effects have physical causes all other causes are excluded – so emergent properties are *epiphenomenal*' (Garrett 2013: 128). Conversely, if emergent properties *are* seen to effect physical change then we will have 'downwards causation' – a thesis which might be thought problematic because of the 'implication … that a scientist cannot do physics completely and adequately without doing biology and, ultimately, psychology!' (Garrett 2013: 150). The severity of these objections is assumed here for the sake of argument.
68  E.g. Wiggins 2005: 476.
69  More broadly, one should say that talking of supervenience one need not necessarily mention 'levels' – see McLaughlin and Bennett 2014.
70  Wiggins 2001: 143.
71  Wiggins 2012: 21. Even Dupré at times seems to express doubts about the existence of emergent properties. While in his 'Spinoza Lectures', he appears to be endorsing a form of causal emergence, elsewhere he assumes a more moderate view, closer to practical, rather than theoretical epistemic anti-reductionism. He writes: 'Perhaps I should concede that everything in the universe supervenes on the total physical state of the universe? Perhaps. But, here, we are so deeply into the domain of speculative metaphysics that I am more than happy to remain agnostic' (2010/2012: 142). This seems to me to be a striking concession on his part – one that is peculiarly over-looked.
72  E.g. in Hare 1952.
73  Dupré 1993: 97.
74  The relations between the emergentist, reductionist and supervenience theses are vexed. While I make no aims to relax them here, it should be noted that the confusion might be taken as a further motivation for pursuing the neo-Aristotelian account offered below (which, insofar as it avoids the puzzles about causality, avoids these more general confusion).
75  Of course, one might talk about 'degrees of reality', but it is not clear that having novel causal properties is a matter of degree, and the disagreement does not seem to be about whether some items have more novel properties than others. (Dupré himself notes the influence of Quine's conception of ontology in the debate, with respect to reductive materialism (1993: 94).)
76  It is helpful to point out that having causal properties correlates to 'ontological commitment' in the Quinean sense, since our best scientific theory must posit bearers of novel causal properties.
77  There is another issue here, worth mentioning in passing. There is more than one type of causal theory and it is not immediately clear which the emergentists are relying on when they defer to Alexander's dictum or whether their position is compatible with them all. For more on the different types of causal theory, see Schaffer 2012.
78  See, for example, Benson 1989: 1067.

79 Dupré, for example, points to methylation (as was seen above) and to the susceptibility of 'lower ranking' Macaque monkeys to cocaine addiction, to ground the claim that novel properties at the social level effect changes at the genetic level (2008b/2012: 257).

80 An entity is 'a temporary eddy in a flux of change'. This quotation comes from Dupré's contribution to the *Auxiliary Hypotheses* blog (2014). The position he describes there is clearing gaining momentum in the philosophy of biology literature. Pradeu has endorsed forms of it, as has Paul Griffiths and Dan Nicholson (many of these views were aired at the 'Process Philosophy' conference that took place in Exeter, in November 2014). (Pradeu's endorsement comes at the end of his introduction to the collection *Individuals Across the Sciences* (forthcoming with Oxford University Press)). It stands in contrast to the typical assumption, voiced by Jack Wilson, that a living entity is 'a finite three-dimensional persisting object' (Wilson 1999: 16).

81 Dupré 2012: 8.

82 Dupré 2012: 99.

83 The distinction between a process ontology and a substance ontology is brought out clearly by Eddy Zemach in 'Four Ontologies' (Zemach 1970).

84 Dupré marks the contradistinction between substance ontologies and process ontologies by stating that the former focuses on *being*, where the latter focuses on *becoming* (Dupré 2014). This rubric is somewhat obscure to me – and for present purposes the point is well made by talking of time-slices and so forth.

85 Stout *forthcoming*.

86 Steward 2015.

87 Steward 2015: 3.

88 'Heraclitus' conceptions of flux, fire and material persistence', (revised version, in *Twelve Essays* (2016): 34).

89 Wiggins 2012: 13–14.

90 Wiggins 2001: 31 In some ways, none of this need concern Dupré (or indeed, Wiggins himself). Both of them, we might say, can have their cake and eat it, since one is eating a process and the other a materially coincident substance. Dupré writes at the end of his addition to the Auxiliary Hypotheses blog: 'As a committed pluralist, I don't want to assert dogmatically that the world is composed of processes not things.' Good. Nor does Wiggins want to claim the converse. Dupré goes on

> However, I am confident that a process ontology provides a generally more illuminating view of the living world, and certainly the same seems true of contemporary physics. At the very least the question whether things or processes provide a better framework for interpreting science is one that should be a central concern of everyone interested in the metaphysics of science.

And again, Wiggins need have no issues with this. The process ontology is borne out within the scientific framework and there is no reason to think that this discredits the entities we pick out in our daily lives. There is room for both things and processes.

## Bibliography

Avise J.C. (2001) 'Evolving Genomic Metaphors: A New Look at the Language of DNA', *Science* 294: 86–87.

Bakhurst, D. (1991) *Consciousness and Revolution in Soviet Philosophy: From the Bolsheviks to Evald Ilyenkov* (Cambridge: Cambridge University Press).

Bakhurst, D. (2005) 'Wiggins on Persons and Human Nature', *Philosophy and Phenomenological Research* 71(2): 462–469.

Bechtel, W. and Richardson, R. (1993) *Discovering Complexity: Decomposition and Localization as Strategies in Scientific Research* (Princeton: Princeton University Press).

Bedau, M. (2003) 'Downward Causation and Autonomy in Weak Emergence', *Principia* 6: 5–50.

Benson, K.R. (1989) 'Biology's "Phoenix": Historical Perspectives on the Importance of the Organism', *American Zoologist* 19(3): 1067–1074.

Boycott, B. (1998) 'John Zachary Young: 18th March 1907–4th July 1997', *Biographical Memoirs of Fellows of the Royal Society* 44: 486–509.

Brigandt, I. and Love, A. (2008) 'Reductionism in Biology', in E.N. Zalta (ed.) *The Stanford Encyclopedia of Philosophy* available online at http://plato.stanford.edu/archives/fall2008/entries/reduction-biology/.

Carnap, R. (1934) *The Unity of Science* (London: Kegan Paul, Trench, Trubner and Co).

Cartwright, N., Cat, J., Fleck, L. and Uebel, T. (1995) *Otto Neurath: Philosophy between Science and Politics* (Cambridge: Cambridge University Press).

Dawkins, R. (2006) *The God Delusion* (New York: Bantam Books).

Driesch, H. (1908) 'The Foundations of the Physiology of Development', 'Experiments on the Egg of the Sea-Urchin', in H. Driesch *The Science and Philosophy of the Organism*, Gifford Lectures for 1907 (London: Adam & Charles Black).

Dupré, J. (1993) *The Disorder of Things: Metaphysical Foundations of the Disunity of Science* (Cambridge, MA: Harvard University Press).

Dupré, J. (2008a/2012) 'The Constituents of Life 1: Species, Microbes, and Genes', in J. Dupré *Processes of Life* (Oxford: Oxford University Press).

Dupré, J. (2008b/2012) 'Against Maladaptationism: Or, What's Wrong with Evolutionary Psychology', in J. Dupré *Processes of Life* (Oxford: Oxford University Press).

Dupré, J. (2010/2012) 'It is not Possible to Reduce Biological Explanations to Explanations in Chemistry and/or Physics', in J. Dupré *Processes of Life* (Oxford: Oxford University Press).

Dupré, J. (2012) *Processes of Life* (Oxford: Oxford University Press).

Dupré, J. (2014) 'A Process Ontology for Biology', on the *Auxiliary Hypotheses* blog, available online at http://thebjps.typepad.com/my-blog/2014/08/a-process-ontology-for-biology-john-dupré.html.

Garrett, B. (2013) 'Vitalism versus Emergent Materialism', in C. Wolfe and S. Normandin (eds) *Vitalism and the Scientific Image in Post-Enlightenment Life Science, 1800–2010* (New York: Springer).

Grene, M. and Depew, D. (2004) *The Philosophy of Biology: An Episodic History* (Cambridge: Cambridge University Press).

Griffiths, P.E. (2001) 'Genetic Information: A Metaphor in Search of a Theory', *Philosophy of Science* 68(3): 394–412.

Haraway, D. (1976) *Crystals, Fabrics, and Fields: Metaphors of Organicism in Twentieth-Century Developmental Biology* (Harvard: Yale University Press).

Hare, R.M. (1952) *The Language of Morals* (Oxford: Oxford University Press).

Hein, H. (1972) 'The Endurance of the Mechanism: Vitalism Controversy', *Journal of the History of Biology* 5(1): 159–188.

Hull, D. (1972) 'Reductionism in Genetics – Biology or Philosophy?' *Philosophy of Science* 39: 491–499.

Kauffman, S. (1971) 'Articulation of Parts Explanations in Biology and the Rational Search for Them', *Boston Studies in the Philosophy of Science* 8: 257–272.

Kim, J. (1992) '"Downward Causation" in Emergentism and Nonreductive Physicalism', in A. Beckermann, H. Flohr and J. Kim (eds) *Emergence or Reduction? Essays on the Prospects of Nonreductive Physicalism* (Berlin and New York: Walter de Gruyter), 119–138.

Kim, J. (1993) 'The Non-Reductivist's Troubles with Mental Causation', in J. Heil and A. Mele (eds) *Mental Causation* (Oxford: Clarendon Press), 188–210.

Lewontin, R. (1993) *The Doctrine of DNA: Biology as Ideology* (London: Penguin Books).

Lovibond, S. (1996) 'Ethical Upbringing: from Connivance to Cognition', in S. Lovibond and S.G. Williams (eds) *Essays for David Wiggins: Identity, Truth and Value* (Oxford: Blackwell Publishing).

Malaterre, C. (2013) 'Life as an Emergent Phenomenon: From an Alternative to Vitalism to an Alternative to Reductionism', in C. Wolfe and S. Normandin (eds) *Vitalism and the Scientific Image in Post-Enlightenment Life Science, 1800–2010* (New York: Springer).

Mayr, E. (2004) *What Makes Biology Unique? Considerations on the Autonomy of a Scientific Discipline* (Cambridge: Cambridge University Press).

McLaughlin, B. and Bennett, K. (2014) 'Supervenience', *The Stanford Encyclopedia of Philosophy*, E.N. Zalta (ed.) *(forthcoming)* available online at http://plato.stanford.edu/archives/spr2014/entries/supervenience/.

Nagel, E. (1961) *The Structure of Science: Problems in the Logic of Scientific Explanation* (New York: Harcourt, Brace & World).

Nagel, T. (1998) 'Reductionism and Anti-reductionism', in G.R. Bock and J.A. Goode (eds) *The Limits of Reductionism in Biology* (Chichester: John Wiley & Sons): 3–10.

Nicholson, D. and Gawne, R. (2013) 'Rethinking Woodger's Legacy in the Philosophy of Biology' *Journal of the History of Biology* (New York: Springer).

Normandin, S. and Wolfe, C.T. (eds) (2013) *Vitalism and the Scientific Image in Post-Enlightenment Life Science, 1800–2010* (New York: Springer).

O'Connor, T. and Wong, Hong Yu (2009) 'Emergent Properties', in E.N. Zalta (ed.) *The Stanford Encyclopedia of Philosophy*, available online at http://plato.stanford.edu/archives/spr2009/entries/properties-emergent/.

Pradeu, T. and Guay, A. (eds) *(forthcoming)* *Individuals Across the Sciences* (Oxford: Oxford University Press).

Putnam, H. (1975) *Mind, Language, and Reality: Philosophical Papers* (Cambridge: Cambridge University Press).

Roe, S.A. (2003) 'The Life Sciences', in R. Porter (ed.) *The Cambridge History of Science Volume 4: Eighteenth-Century Science* (Cambridge: Cambridge University Press): 397–416.

Rose, S., Lewontin, R.C. and Kamin, L.J. (1984) *Not in Our Genes: Biology, Ideology and Human Nature* (London: Penguin Books).

Rosenberg, A. (2003) 'Reductionism in a Historical Science', in M.H.V. van Regenmortel and D.L. Hull (eds) *Reductionism in the Biomedical Sciences* (Oxford: Wiley-Blackwell).

Rosenberg, A. (2006) *Darwinian Reductionism: Or, How to Stop Worrying and Love Molecular Biology* (Chicago: University of Chicago Press).

Sarkar, S. (1992) 'Models of Reduction and Categories of Reductionism', *Synthese* 91: 167–194.

Salmon, W. (1989) *Four Decades of Scientific Explanation* (Minneapolis: University of Minnesota Press).

Schaffer, J. (2003) 'Is There a Fundamental Level?', *Nôus* 37(3): 498–517.

Schaffer, J. (2012) 'Causal Contextualism', in M. Blaauw (ed.) *Contrastivism in Philosophy* (London: Routledge): 35–63.

Scriven, M. (1962) 'Explanations, Predictions, and Laws', in H. Feigl and G. Maxwell (eds) *Minnesota Studies in the Philosophy of Science*, vol. 3 *Scientific Explanation, Space, and Time* (Minneapolis: University of Minnesota Press), 170–230.

Smith, J.S. (2000) 'The Concept of Information in Biology', *Philosophy of Science* 67(2): 177–194.

Steward, H. (2015) 'What is a Continuant?', *Aristotelian Society Supplementary Volume* 89(1): 109–123.

Stout, R. (*forthcoming*) 'The Category of Occurrent Continuants', *Mind*.

Taylor, C. (2003) 'Ethics and Ontology', *The Journal of Philosophy* 100(6): 305–320.

Theurer, K.L. (*forthcoming*) 'Seventeenth-century Mechanism: An Alternative Framework for Reductionism', in *Philosophy of Science Association 23rd Biennial Meeting* (San Diego, CA).

Toulmin, S. and Goodfield, J. (1962) *The Architecture of Matter* (Chicago: University of Chicago Press).

Waters, C.K (1990) 'Why the Antireductionist Consensus Won't Survive the Case of Classical Mendelian Genetics', in A. Fine, M. Forbes and L. Wessels (eds) *Proceedings of the biennial meeting of the Philosophy of Science Association* 1 (East Lansing, MI: Philosophy of Science Association): 125–139.

Wiggins, D. (1980) *Sameness and Substance* (Cambridge, MA: Harvard University Press).

Wiggins, D. (1987/1991) *Needs, Values, Truth* (Oxford: Blackwell).

Wiggins, D. (1997) 'Languages as Social Objects', *Philosophy* 72(282): 499–524.

Wiggins, D. (2001) *Sameness and Substance Renewed* (Cambridge: Cambridge University Press).

Wiggins, D. (2005) 'Reply to Bakhurst', *Philosophy and Phenomenological Research* 17(2): 442–448.

Wiggins, D. (2012) 'Identity, Individuation and Substance', *European Journal of Philosophy* 20(1): 1–25.

Wiggins, D. (2016) *Twelve Essays* (Oxford: Oxford University Press).

Wilson, E.O. (1975) *Sociobiology: The New Synthesis* (Cambridge, MA: Harvard University Press).

Wilson, J. (1999) *Biological Individuality* (Cambridge: Cambridge University Press).

Wimsatt, W. (1976) 'Reductive Explanation: A Functional Account', in R.S. Cohen and A. Michalos (eds) *Proceedings of the 1974 meeting of the Philosophy of Science Association* (Dordrecht: D. Reidel): 671–710.

Wimsatt, W. (1980) 'Reductionistic Research Strategies and their Biases in the Units of Selection Controversy', in T. Nickles (ed.) *Scientific Discovery: Case Studies* (Dordrecht: D. Reidel): 213–259.

Woodger, J.H. (1930–1931) 'The "Concept of Organism" and the Relation Between Embryology and Genetics', *Quarterly Review of Biology* 5: 1–22.

Woodger, J.H. (1937) *The Axiomatic Method in Biology* (Cambridge: Cambridge University Press).

Woodger, J.H. (1952) *Biology and Language* (Cambridge: Cambridge University Press).

Wolfe, C.T. (2010) 'Do Organisms Have an Ontological Status', *History and Philosophy of the Life Sciences* 32(2): 195–231.

Wright, R. (1994) *The Moral Animal: Evolutionary Psychology and Everyday Life* (New York: Pantheon Books).

Young, J.Z. (1971) *An Introduction to the Study of Man* (Oxford: The Clarendon Press).

Zemach, E. (1970) 'Four Ontologies', *The Journal of Philosophy* 67(8): 231–247.

# 8  Aristotelian organisms

Wiggins wants to hold both that everyday organisms are real and that they are more than the ontological run-off of their microscopic parts. The emergentist's route is open to him, but there is another way which I take to be more in keeping with his descriptivist sensibilities. He may draw more extensively than he has been wont to do on the full metaphysics of *substance*. Positioning himself outside the science-led metaphysical mainstream he can avail himself of the riches of the Aristotelian tradition and extend his analysis of *organisms* (natural substances), the structure of that concept, and the persistence conditions of these beings. Having done so, he might (as suggested in Chapter 6) marry this picture to the physiological-functional account of organismic individuation. In this way, I think, he can answer most of the questions about the identity of persons that vex present-day philosophy.

What follows is a recommendation for Wiggins, rather than an interpretation of his texts. I shall begin with the Kantian thought that our conceptual framework necessarily partitions the biological realm into *genuine unities*. To some – including Wiggins – this may seem an odd starting point, but I hope by the end of this chapter to have shown how Kant's analysis complements the methodological approach we find in *S&SR* and enriches the picture of substance we find there. My advocacy of organisms as genuine unities will have implications for how their parts are conceived. It will involve understanding an organism's parts by reference to the whole. The *organism* concept (I shall say) is an intrinsically *anti-reductive* one. In the second section I shall join this idea to the Aristotelian or neo-Aristotelian ideas about ontological *priority* and the causal power of organisms (thoughts which appear in *ISTC*, but which – if not explicitly abandoned – have since faded into the background of Wiggins's work). At the end of the chapter I return to the explication of the persistence conditions of human beings.

## 1  Organic unities

Let us build first upon the unlikely intellectual connection announced in Chapter 1 between Wiggins and Immanuel Kant. For Wiggins, human beings – human organisms – are things that we cannot avoid picking out in our day-to-day existence. This thought is found too – famously – in Kant's discussion in *Critique of*

*Judgement.*[1] In that work Kant also emphasizes how hard it is to avoid understanding these things in *teleological* terms. Teleological language is ineliminable from the life sciences, he claims:

> [I]t is absurd for human beings ... to hope that perhaps some day another Newton might arise who would explain to us, in terms of natural laws unordered by intention, how even a mere blade of grass is produced.[2]

There can be no such Newton – Kant thinks – because natural entities resist the explanations by mechanical principles exemplified by Newtonian physics. This is to say that the nature and behaviour of the entities we find in the biological realm (organisms) cannot be accounted for by reference to the presence and behaviour of their constituents. They must be seen as *more* than the sum of their parts. We conceive of them as final ends to which the functioning of their parts are directed. For Kant, on the interpretation offered by Philippe Huneman (and others[3])

> [t]hese entities display a specific relationship between wholes and parts, in which the parts are seen to presuppose the whole in order to be accounted for.[4]

In our pre-theoretical thinking, the organism is conceptualized as 'prior' to its parts. As Kant sees these matters, we understand the *whole* before we can set to work understanding the form and function of its assorted appendages. This seems to me to be exactly how we approach things that belong to the biological realm. Think how difficult it would be for a child to ask how noses work if she had not already grasped how and where the noses belong in the human physiognomy. Contrast that with the child's asking how light-bulbs work. She might care only about the source of some beguiling radiance, and need not even recognise the larger machine (the night-light, the traffic-light, the head-light, etc.) of which it is a part. In other words, there are things that we have to pick out *before* we can start understanding their parts. In our everyday lives, we single out organic *wholes* and only then make the imaginative leap to think of their parts in the abstract. Here, roughly speaking, is what it is to be a genuine unity in the Kantian sense.

For Kant, however, as for many who give short shrift to conceptual analysis, this is an epistemological fact with no ontological bite. It may be that we apprehend organisms as unities and conceive of their parts *as* parts of the whole. But, as Huneman notes, Kant disclaims any theological – or, more broadly, metaphysical – commitments.[5] Picking out organisms is something we do, but not 'a statement of the real'.[6] The organism is a 'heuristic fiction' (a *regulative* principle, in the Kantian idiolect, and not a *constitutive* one).[7]

Not everyone, however, has been so quick to dismiss the metaphysical relevance of this analysis. We might look for example towards recent work of Charles T. Wolfe, who demonstrates how Kant's account may be used for phenomenological ends. In his essay, 'Do Organisms Have an Ontological Status?'[8] Wolfe traces an important connection between Kant's view of the ineliminibility of teleological language and the more recent Darwinian musings of Daniel Dennett and the neuropsychologist, Kurt Goldstein.[9] As Wolfe reads them, Dennett and Goldstein hold that the ability (on which our survival depends) to predict the behaviour of others relies on our treating them as *unities*, and not, say, as a mass of molecules.[10] Wolfe writes:

> [O]ur cognitive or perceptual make-up is necessarily 'organismic' [i.e. it is part of our conceptual framework that we pick out organic unities], and indeed, its being so contributes to our aptitude for survival.[11]

He goes on:

> The projective view, which I have attributed to Kant, Goldstein (in one of his moods) and Dennett, holds that organism is something we project onto the world, a kind of construction of intelligibility … the organism must be treated as an individuality.[12]

While building on Kant's view, Wolfe's conclusion is not the same as Kant's. For he finds a metaphysical moral in the original thought. The organism is a projection, yet '[t]hat we are, by dint of our nervous systems, "projectors" does not mean we project any structure we choose onto the world'.[13] As humans, with a specific conceptual framework, we bring something to bear on the world – but doing so does not mean we necessarily create the structure we find within it.

Without having to reconstruct the arguments that lead Wolfe to this point it will be clear in the light of the previous chapters (see especially Chapter 1) that this is a trajectory Wiggins may well shadow.[14] Organic *unities*, as Kant, Leibniz,[15] Goldstein, Claude Bernard,[16] Wolfe and Aristotle (etc.) all maintain, are a central part of our conceptual framework. For whatever (evolutionary) reason, we partition the natural world into beings that are seen to be more than a particular configuration of their parts. The biological realm might be articulated differently in other frameworks, but this – as Wiggins points out – is not to say that our pre-theoretical description lacks nomological foundation. (Indeed, our human ability to pick out these unities is a scientific fact that may well be accounted for in some 'ultimate' theory.) To refer back to Wiggins's conceptualist-realist phrasing, the fact that we *construe* reality in a certain way does not imply that we *construct* it.

## 2  Aristotelian biology

It is time, if not somewhat passed it, for an overview of Aristotelian metaphysics. I noted in the introduction that Aristotle has been a constant presence in Wiggins's work – and Wiggins is one of his staunchest supporters in contemporary Anglophone philosophy. It is not, then, too far a stretch to suggest that his analysis of *substance* might be extended along Aristotelian lines.

Wiggins will not want to endorse the doctrine of *hylomorphism* (explained below) but he may well be amenable to the notion of *priority*, which it is the aim of the section that follows to furnish.

For Aristotle (as for the descriptivists),[17] the objects of our everyday experience – cats, flowers, human beings – are the primary focus of metaphysical inquiry. In the works collected in the *Metaphysics*, he delves into the underlying structure of these things,[18] and it is in those texts that we find the infamous *hylomorphic distinction*. In the realm of being,[19] Aristotle says, the individuals we encounter in the world are not metaphysically simple but comprised of two distinct elements: matter (*hulê*) and form (*morphê*). All individual things are hylomorphic composites.[20]

This position is elaborated in *Metaphysics* Z by reference to the example of a bronze statue. The bronze statue has a *material* (hylic) aspect; it is made out of matter, in this case, bronze. It also has a *formal* (morphic) one; it is moulded into a certain shape or 'form' (for example, Judith and Holofernes). The form cannot exist without being 'enmattered'; that is, *realized* in matter. The matter may exist without the form (the statue may be destroyed when the bronze survives). Uninformed matter is not a distinct particular but some (relatively) undifferentiated mass. It is the *form* that determines individuals – as the *form* of a statue is what makes it the kind of thing that we can individuate.

The statue paradigm is slightly misleading. Form is more than a particular *shape* into which something is moulded[21] (the statue of Judith is not a human being). To get a better grip on this consider a more complex artefact. A hatchet, say. Here, Aristotle associates 'form' with a specific *function*; for example, the chopping of wood.[22] In this instance, form is not just the *shape* of a thing. To specify the form is to say in what particular way it works or functions. It is to specify the thing's *mode of being*. So the form of a hatchet is inseparable from its use, to divide wood by chopping through it. The form of a house is to give shelter by means of a roof. The form of the roof is to prevent a deluge by being an impermeable surface.[23]

Furthermore, it is central to Aristotle's picture that a substance's matter must be *of a specific sort*, or of a certain material configuration, such that it can *subserve* the object's function:

> [T]here is a necessity that the axe be hard, since one must cut with it, and if hard that it be of bronze or iron.[24]

That is to say that the matter which will constitute the hatchet must have the right *dispositional properties* for chopping.[25] The matter must have properties that make it amenable to – disposed towards – such activities. Or, to put it in the Aristotelian idiom, the matter must be *potentially* (*dunamei*) what the axe, spade, chisel, will be in *actuality* (*energeia*, *entelecheia*).

What about organisms? In expounding his thoughts about these, Aristotle still relies in part upon an analogy with what it takes to explain the being of artefacts. The organic entities we find around us are matter 'informed' by specific *morphe*.

An animal is matter informed by a specific mode of being[26] just as the hatchet is matter formed to serve a particular function. This natural form, the *psuche*, encapsulates the typical growth and development, the behaviour, of that kind of animal. As Montgomery Furth and L.A. Kosman put it, it encapsulates the animal's 'lifestyle':

> The specific form of an animal's being is … as we might say, its *lifestyle*; what it eats, how it gets food, where it lives, the manner of its reproduction, sensation, movement, etc.: the entire complex of characteristic activities, in other words, that constitutes the manner of its *bios*.[27]

Thus, in describing the metaphysical character of human beings, Aristotle holds that they are entities comprised of *psuche* and matter (matter that is suitable for the subservience of that form).[28] Here is the stem from which Wiggins's metaphysics grows. There are issues about how his interpretations of Aristotle fluctuate,[29] but it should be evident from this analysis, and the previous chapters, that Aristotle's doctrines of *psuche* and *phusis*[30] are among the conceptual ancestors of Wiggins's 'principle of activity'.[31] The connection is made explicit in 'Identity, Individuation, and Substance',[32] where he presents his project in these distinctly Aristotelian terms, describing how natural kinds have a *phusis* or a nature:

> The *phusis* of a thing is its mode of being. It is the principle of activity of a kind whose members share and possess in themselves a distinctive source of development and change.[33]

There is an etymological irony here that would be a shame to pass over. Wiggins's approach is often contrasted with the 'psychological' approach to personal identity. But follow the derivation of that term and one finds Aristotelian *psuche*. In an ancient sense, Wiggins is a psychological theorist, seeing us to survive where our *psuche* continues. His is a modern retelling of the Aristotelian story. Now the question arises: how true does he want to stay to the original? The next section deals with a pressing problem for those who endorse the hylomorphic picture: its seemingly inextricable connection with *final causality*.

## 3 Teleology

Central to the Aristotelian schema is the thought that the *psuche*, the form of the living body, has *causal* powers over the material parts of the organism. As Marjorie Grene puts it, *psuche* is the 'organizing' principle, which arranges the material parts, the elemental compounds, into the organismic (that is, 'organized') structure.[34] In Aristotle, this position is articulated in contrast with the framework outlined by Empedocles.[35] Both Aristotle and Empedocles see matter as split between the four elements: water, wind, earth and fire. Both philosophers understand these elements to have their own elemental natures (which is why uninformed matter may have dispositional properties) and to be mixable into compounds. Where they differ, however, is in the causal power that they attribute to these four fundamental stuffs.

For Empedocles, everything in the earthly realm arises from the interaction between the elements and their elemental natures. Moved together by their environments,[36] they mix and, through chance and necessity, form more ordered compounds – mud, bark, bronze, blood, etc. – which themselves mix and create even more complex higher level structures.[37] The reductionist logic of this process is exemplified most strikingly in the Empedoclean theory of evolution; he describes a stage in cosmic history where heads and trunks and limbs are formed from the elements, and roll about the primordial tundra, occasionally bumping into each other and combining to produce bizarre, chimerical, 'scrambled animals', such as ox-headed men and human-faced cows (some of which form viable combinations, which then survive).[38] In the Empedoclean universe, one thinks of organisms as having been originally caused by elements combining in certain conditions. Ultimately, the organism is seen to be the product of a causal chain *reaching up* from these elements, the most fundamental constituents of reality.

This conception of a uni-directional causal ascent stands in marked contrast to the Aristotelian model of biological systems. Aristotle's biology is open to divergent interpretations, but whether or not it is credited with an anachronistic vitalism,[39] he explicitly opposes the strict reductionist picture that we find in Empedocles. Specifically, he views the *psuche*, the living activity of a creature, as playing a causal role that is *not* an extension of some upwards-reaching causal chain.[40] Rather, he sees *psuche* itself as having causal influence over its parts: the causal chain does not move upwards, emerging mechanically from the nature of the elemental constituents. Rather, as Furth puts it, it 'reaches down' into the elemental miasma, and thereby gives it structure.[41]

As will be clear from the overview of the reductionist/emergentist debate, the echoes of the arguments of Empedocles and Aristotle are still ringing. Yet Aristotle's causal anti-reductionism is not exactly the emergentist's one. For Aristotle, the organism forms the parts, as the end-product, the *telos*, to which the organization of those parts is directed. In the emergentist's picture, the novel causal power is not a function of the fixed end-point of a biological process. This is to say that Aristotle's hylomorphism is wedded to the doctrine of *teleological causality*.[42]

According to Aristotle a 'teleological' cause is a goal to be achieved. The case of artefacts furnishes an analogy. Consider an intentional action – your desire to hammer some nails, for instance. Hammering nails is your goal. To achieve this goal you assemble a lump of wood and a sharp bit of iron and fashion a hammer. In some sense, then, the *goal* has caused the organization of the wood and iron to be thus and so. Your notion of the activity to be performed has organized your effort to organize matter in a specific way.

Nowadays, few people will find this account of causation plausible when it is transferred from artefact to organism (especially where, like Aristotle, they do not posit any kind of purposive action by a creator God).[43] Aristotle gives numerous examples where natural processes are construed as being *caused* by the organism's 'living well'; a plant's roots push down for nutrition,[44] or its leaves grow to protect its fruit.[45] The life of the organism is seen as the end that stands as the primary cause of organic development; the movements of the elements are all *directed* towards the growth and persistence of a living being.

This is all of a piece with the Aristotelian conception of *dunamis* and *energeia*, mentioned above. For the organic form/lifestyle to be actualized, the matter that constitutes it must be of the right sort to *support* the characteristic activity (*energeia*). So it must be potentially living. And for it to have the capacity (*dunamis*) to support a particular sort of *bios*, the matter must be structured as, for example, organs that perform the required functions of that type (for example, if the individual is a bipedal thing then two legs are required for locomotion, lungs for breathing, etc.). And for those organs to function in this way, they need in turn to be made of certain matter (matter which has the potential to so function), and onwards, *downwards* to the Empedoclean elements. In short, the 'lifestyle' of the organism – how it gets food, where it lives, etc. – arranges matter into a structure that can realize it.

To those living in a post-Darwinian age the idea that a plant, say, is organized in a certain way *because* being so organized is good for it,[46] will seem both mysterious and obsolete,[47] since we can much better explain organic arrangement as, for example, the outcome of non-purposive mutations that allow organisms (and, consequently, those particular traits) to survive. The idea that inanimate things 'seek' natural end-states sounds profoundly strange to modern ears. This is not the territory that Wiggins inhabits.[48] He writes, dryly:

The failure of Aristotelian science is final.[49]

He does however add that

[i]ts failure does not entail ... that science has shown that every explanation in any way worth having of anything worth knowing must eventually find expression at the level of the science that has displaced the human world view.[50]

Like Aristotle, Wiggins sees organisms to be substances *par excellence*. For Aristotle, this notion is tied to the thought that such things are *unities*, and the discussion above presents one strand of argument for why he thinks this. Living things are unified (and thus more clearly exemplify the category of substance) because they have causal power over their parts. In a way, this complements – though fails to coincide with – the emergentist's picture, in which the causal standing of a thing is taken to correspond to its metaphysical status. Yet Wiggins's use of Aristotle does not rest on the teleological power of the organic whole. In the next section it is argued that his interests lie with another element of Aristotle's metaphysics: *ontological dependence*.

## 4 Ontological dependence

Despite distancing himself from the teleological aspects of Aristotelian science, Wiggins finds Aristotle's picture useful. He takes it to elucidate our everyday *substance* concept and to describe the metaphysical make-up of the continuants we are constantly picking out. I have already described how some philosophers think that organisms must be understood as unities, and it is argued below that Wiggins can appeal to Aristotle's account in explaining this. In addition to the difficult claims about causation, there is another strand of Aristotelian anti-reductionist thought which intersects neatly with the descriptivist's project. Roughly put, an organism is a genuine unity because the existence of the parts *ontologically depends* (in a conceptual, non-causal way) on the whole.

'Ontological dependence' was introduced, briefly, in Chapter 3. We can turn to Fabrice Correia, for a slightly more nuanced assessment. He describes it as

> a term of philosophical jargon which stands for a non-well delineated, rich family of properties and relations which are usually taken to be among the most fundamental ontological properties and relations.... A dependent object, so the thought goes, is an object whose ontological profile, e.g. its existence or its being the object that it is, is somehow derivative upon facts of certain sorts – be they facts about particular objects or not.[51]

In discussing the metaphysical character (or 'ontological profile') of *substances*, Wiggins deploys notions of ontological dependence (or 'posteriority') that come from the Aristotelian tradition (this was seen at the end of Chapter 3). Properties *depend on* the substances in which they inhere (this is how he articulates the metaphysical relevance of the subject/predicate distinction). 'Real unity' was also mentioned as an aspect of the pre-theoretical *substance* concept and this, too, was seen to be explicable in terms of dependence. In cases of genuine or real unity, the parts of an item are understood to be *ontologically dependent on* the whole. The time has come to present a more precise rendering of this Aristotelian claim. In *ISTC*, in a passage since faded into obscurity, Wiggins offers an introduction:

> Now the reason why the material parts of *x* are accounted posterior to *x* by Aristotle seems to be something like this. Suppose we take the example of parts of the body. They have to be picked out or individuated in some way or other, and any correct way of picking them out will have to make clear *what* exactly we are picking out. But this involves making clear the existence and persistence conditions (for Aristotle slightly peculiar) of the bodily parts we do pick out.... These can only be entirely correctly given if we pick these parts out *as parts of this or that living body*. (That anyway is Aristotle's view of bodily parts. For him such are really living-bodily-parts. The generalizable point is that the picking out of parts *p* must *somehow* make clear what *p* are ...) So Aristotle writes:

And the finger is defined by the whole body. For a finger is a particular kind of part of a man. Thus such parts are material, and into which the whole is resolved as into matter, are posterior to the whole; but such as are the parts in the sense of parts of the formula and of the essence as expressed in the formula [*tou logou kai tes ousias tes kata ton logon*], are prior. Either all or some of them.

(*Metaphysics* 1035$^b$ following, trans. Tredennick)[52]

Wiggins is describing a form of *essential dependence*. It is part of the essence of a finger – hand, organ, etc. ... – to be a part of an organism. Put slightly differently, the statement of the essence (the 'logos' or 'real definition') of that thing will make reference to the whole of which it is a part (or the whole is a 'constituent' of that real definition). Kathrin Koslicki, in 'Substance, Independence and Unity', has done much to make this clearer, and describes the Aristotelian model of ontological dependence in the following terms:

An entity, Φ, ontologically depends on an entity (or entities), Ψ, just in case Ψ is a constituent (or are constituents) in a real definition of Φ.[53]

An illustration may help here. Consider Aristotle's various discussions of the essence of *eyes*. Where body parts are concerned, Aristotle takes their essence to relate to their *function* – this is the thought captured in the *Meteorology*, where he writes:

What a thing is is always determined by its function: a thing really is itself when it can perform its function; an eye, for instance, when it can see.[54]

Aristotle, note, is not stating that the eye must always be *performing* its function to be an eye. That would be a strange claim indeed. Rather, it must be such that it *can* perform its function. For an eye to be an eye it must have the potential (*dunamis*) to see (this is part of what Koslicki calls its 'real definition'); our eyes do not cease to be eyes when we sleep, since they retain that potential to function (though it remains unactualized). The crucial point is that eyes cannot properly be said to see by themselves. *People* see. The function an eye performs is the function for the whole, so an eye must be integrated into an organized whole for it to have the capacity to perform its essential function. The eye, like the finger, is 'defined by the whole body'; the real definition of an eye essentially refers to the organism of which it is a part. So can eyes exist separately from the whole?

Evidently, even of the things that are thought to be substances, most are only potentialities – e.g. the parts of animals (for none of them exists separately, and when they *are* separated, then they too exist, all of them, merely as matter).[55]

For Aristotle, when an organ is separated from the organism, it is only an organ 'homonymously'.[56] That is, though the word 'eye' is used equally to refer to the living eye in my head, and to the detached and decomposing eye on the surgical plate, 'the definition of being which corresponds to the name is different'.[57] One exists as a part of a whole, with a particular function – the other 'merely as matter'.

> [W]hen seeing is removed the eye is no longer the eye, except in name – no more than the eye of a statue or of a painted figure.[58]

In an age where organ transplantation has become almost routine in the West, it might seem odd to suggest that an eye separated from a human body (or a heart so separated, or a liver, or whatever) ceases to be an organ. These concerns are addressed in the next chapter, but it is worth mentioning that the animalists, among others, are generally sympathetic to Aristotle's position. We find the claim, in Eric Olson's *What Are We?*, that 'there are no brains or heads or upper halves'.[59] More specifically, 'there are no *undetached* brains, heads, and so on'.[60] This relates to his endorsement of *biological minimalism* (the thesis that only living things and mere-ological simples exist), which, according to Peter van Inwagen, is a direct descend-ent of Aristotle's substantialist view.[61] I do not intend to assess the tenability of that programme here, but I think that it is more than a little interesting that the Aristote-lian view of organs is regaining currency.[62]

The central thought in this section is that the homonymy picture is not the result of an argument exclusively focused on causal dependence. There are two sides to Aristotle's anti-reductionism. On the one hand he claims that organisms are genuine unities because of the *causal power* they have over their parts. But on the other, he explains their substancehood in terms of *priority* and *posterior-ity*: organisms are real unities because their parts are *ontologically dependent* upon them. And while Aristotle interweaves these strands, the suggestion here is that they need not necessarily be so woven.

Wiggins does not associate himself with Aristotelian final causality. Rightly so. Yet he may, if he follows this recommendation, find in Aristotle a way of capturing our thoughts about organic unity without invoking that problematic principle. The causal story and the claims about dependence are *separable*. These issues are not approached head on in Wiggins's texts, but he is doubt-lessly aware of them, and we find something very close to a statement of this separability in his paper 'Substance':

> The extent that anything is not *in* other things ... it enjoys a certain auto-nomy. Something that has this autonomy may be causally dependent on other things in the way in which the infant depends on the mother; but onto-logically speaking, it is still independent.[63]

Just because something is causally dependent on something else does not mean it is *ontologically* dependent on that thing. Nor, the thought goes, are causal dependence relations the only relations of metaphysical import: this kind of ontological-cum-conceptual dependence is equally worthy of attention. Again, if Wiggins were to think this, he would not be alone in doing so. In addition to contemporary discussions (like those of Koslicki, Kit Fine and Jonathan Schaffer) ontological dependence is a relation in notably good standing in the descriptivist tradition. We find it in Strawson's work, where he explicitly relates ontological priority to individuation, in a way that is certainly amenable to Wiggins's texts. For Strawson, a thing *x* is ontologically prior to another thing *y* if and only if it would be impossible for us to identify *y* unless we could identify *x*, but not vice versa.[64] Applying this to the case of organs, we find again something close to Aristotle's homonymy view: we cannot identify a heart as a heart without thinking of its role in the organic system – but we can, and often do, identify organisms without picking out hearts. Attention to our everyday thoughts and individuative practices shows a certain hierarchical dependence of organs on the organisms of which they are a part.

   The proposal, then, is that Wiggins can take Aristotle's discussion of dependence to enrich our understanding of a central element of our conceptual framework.[65] In our pre-theoretical experience we treat some things – including organisms – as possessing principles of activity, as being subjects of predication, and as being *real unities*. This latter aspect of substancehood – where an item is seen to be 'more than the sum of its parts' – is captured well by Aristotle's thoughts about dependence. An item is a real unity when its parts are *ontologically* (but not necessarily *causally*) dependent upon it, i.e. when the capacity to perform their essential function depends on their being integrated into an organized whole. The result is a clearer picture of an entity that is poorly captured in the 'scientific' framework. We started this chapter by wondering how Wiggins could honour the structure of pre-theory, eschew biological council and simultaneously enrich our understanding of our metaphysical nature – the analysis of our practical idea of an *organism* has offered an answer.

## 5  Persistence conditions of organic substances

Let us apply all this to the question of the identity, unity and persistence of organic substances. First of all, these claims about ontological dependence suggest that we must take seriously the thought that we cannot conceive of organs (the parts of natural substances) as *separate* from the organized whole – and that this conceptual fact has metaphysical importance. This Aristotelian thought[66] is also echoed by Hegel – a philosopher who makes a notable, but over-looked, epigrammatic appearance in *S&S*:[67]

> The limbs and organs for instance, of an organic body are not merely parts of it: it is only in their unity that they are what they are.[68]

This is less about *our* persistence conditions and more about the persistence conditions of our parts – but the latter is clearly linked to the former. If Wiggins elucidates our pre-theoretical *substance* concept along the suggested lines then he will hold that organisms, as natural substances, cannot conceivably be dismantled. That is, it is beyond our conceptual limits to think of them as suffering *disassembly*. Natural substances are ontologically *prior* to their parts. They cannot, then, be separated *into* their parts because their parts can only exist in the integrated unity. This is not to say that other biological individuals (genetic individuals or immunological individuals) cannot be broken into bits; it is to say that an organism, where 'organism' features now as a specific technical term, cannot be dismantled.

Nor is this to say that organisms cannot lose bits of themselves. We lose hairs, teeth and sometimes, unfortunately, larger body parts like fingers and legs. But, once separated, the thought goes, these things cease to be capable of realising their essential function and thus cease to exist (a detached 'finger' is not a finger but a material aggregate with the shape of a body part); as a result, the organism becomes smaller (rather than spatially discontinuous).[69]

A finger is one thing, but a heart is another. The one assists the organic system, the other is systemically central. In the normal course of events, I would die almost instantaneously if my heart were taken from my chest. The same would be true of the removal of my brain, or my lungs, or my stomach. Transplantation medicine – the subject of the next chapter – presents interesting cases where body parts such as these can be taken out and *reinserted* into the body. The logic and practice of transplantation suggests that organisms *can* be dismantled and rebuilt, and this is a claim that I believe a *substantialist* metaphysician will want to resist. As was hinted at in Chapter 3, the disassemblage of artefacts is unproblematic because of the relative independence of artefactual parts from artefactual wholes. A watch can survive in a disassembled state because its parts can be seen to exist when separated from one another. The same is not true for the organic substance. It will be argued that, strengthening our conception of organic persistence conditions, Wiggins could present a principled response to Sydney Shoemaker's infamous story of human brain transplantation.

\*   \*   \*

In conclusion let me draw together some loose ends concerning coincident entities, the divide between pre-theory and science, and the possibility – critically examined in Chapter 6 – of deferring to some scientifically informed theory in order to flesh out the *human being* principle.

The first chapter of this book aimed to show that the descriptivist's project is at the very least plausible. It is possible to question the univocity of science-led metaphysics.[70] The scientific perspective might capture certain truths; it need not capture them all. There are grounds to think that we should keep both eyes open, metaphysically speaking – the living eye of pre-theory and the a-temporal eye of science – in order to see the world in all its depths and dimensions.

Wiggins focuses on substances, but he insists that other things – quarks, processes, etc. – inhabit the world alongside them. What entities we pick out depends upon the framework we are working within. As was laid out in Chapter 1, there are no obvious grounds for saying that one framework has priority over another. Reality may be populated by metaphysically various entities. We need not be restricted to those entities captured by science. There are others besides. Among them are the human organisms of everyday experience, which Wiggins takes as his subject and which stand on metaphysically equal terms to the processual beings and the material aggregates described by emergentists and reductionists.

The advantages of this open-minded realism have been outlined, alongside some of its more troublesome features. The most significant concern was that, read thus, Wiggins's project seems increasingly detached from the more detailed picture of reality offered by science. With all this talk of contrasting frameworks and beings, it appeared a wedge had been driven between pre-theory and science, and the issue that rose to increasing prominence was whether the entities found in one framework could be discussed, analysed and articulated in another. We started Chapter 6 with the claim – now much diminished in plausibility – that 'human beings' are easier to individuate than persons. The focus of this chapter has been to regain some of this lost ground.

Imagining that the gulf between 'pre-theory' and 'science' was too broad to cross, I have suggested that we can examine our *conceptual framework* in the hope of clarifying our persistence conditions. This is the deflationary response available to Wiggins. One suspends judgement about the possibility of communication between frameworks and makes a start on answering questions about survival through descriptive means. We see ourselves as substances and there are some changes which consideration of that concept suggests we simply cannot undergo (disassembly being one).

But in saying this I do not mean that we should relinquish the insights offered to us by biology. We have occasion now to revisit the poorly formed thought found at the end of Chapter 6: is the strait between 'science' and 'pre-theory' really so impassable? First of all, the critical reader will no doubt be aware of the rather uncritical use of the term 'science' in the preceding pages; it has been taken to include any number of sometimes contradictory disciplines – and it is a bold, if not bald, claim to say that *all* of them clash with pre-theory (this is something we will return to at the end of Chapter 9).[71] I now surrender this claim.

Recall the 'physiological-functional' account of the organism. This we saw as conforming closely to everyday, phenomenal individuation. At first, it seemed insufficient to our purposes; it was too heavily reliant on the phenomenal picture to justify it.[72] Unlike the genetic view, it did not appear to latch onto 'robust patterns' in nature. It offered no stable grounds for demarcating biological individuals. (No wonder other theorists have tried to supplement it with principles of inclusion, such as the immunological one.) Now, however, given the work in the intervening sections, I think we can say that there is independent justification to mark out the phenomenal individual. The phenomenal picture itself is not without metaphysical standing. Physiology, then, and functional accounts of the organism, might consequently be used to flesh out the ordinary model of the human, which sustains them. There is a more scientifically precise way of understanding organic substances. Geneticists may not be able to help us, but physicians might – if we allow theory and pre-theory to combine to elucidate 'the characteristic development, the typical history, the limits of any possible development or history, and the characteristic mode of activity' of human beings.[73]

When it comes to the puzzle cases which troubled the physiological-functional view, the descriptivist now has the means to respond to them. Are endosymbionts parts of us? If he follows the analysis above, Wiggins will deny that they are.[74] We do not see ourselves, pre-theoretically, to be heterogeneous entities. This response is not a knee-jerk 'intuition'; it emerges out of a study of our pre-theoretical concept of *substance*. A substance is an entity upon which its parts ontologically depend. In a statement of the essence of a heart there is a necessary reference to the role it plays in the functioning of the whole. Compare this to a statement of the essence of the bacteria with whom we enjoy a symbiotic relationship; they need not be understood as playing roles in the organic whole. They do not ontologically depend, in the sense explained above, on the human being that hosts them.

Analysis of pre-theory can aid our understanding of the targets of our biological investigation. This approach allows Wiggins to extrapolate more clearly our distinctive principle of activity. It is fully in line with his descriptivist commitments. Furthermore, this approach gives him an advantage over his nearest competitors in the personal identity debate. He has resources that the science-led animalists lack: how might *they* bolster the physiological-functional account? It remains a live question for Olson and Snowdon.

In the next chapter it is suggested that this reading provides a novel response to a particular puzzle of the personal identity debate. We see, in Sydney Shoemaker's infamous brain transplantation narrative, a shift in metaphysical focus from one type of entity onto another, from a *substance* onto *a biological artefact*. We will find, then, the fulfilment of the proposal offered above: that a descriptive analysis of *substance* (combined with the physiological-functional account) can help us rule on questions about personal identity.

## Notes

1 Kant 1790/1987.
2 Kant 1790/1987 §75: 282–283 Did Darwin fill this role? There is not the space to discuss this in the depth it deserves, but the profusion of teleological language in Darwin's work suggests that, although he explained it differently, he still conceived the natural world in some way teleologically.
3 See, for example,Wolfe 2010, Ginsborg 2001.
4 Huneman 2007: 5–6.
5 Huneman 2007: 5.
6 Wolfe 2010: 21.
7 Wolfe 2010: 19.
8 Wolfe 2010.
9 Dennett 1987.

10 Wolfe 2010: 24.
11 Wolfe 2010: 21.
12 Wolfe 2010: 21.
13 Wolfe 2010: 29.
14 Here again we see the significant similarities between the descriptivist tradition and the phenomenological one.
15 Leibniz 1695/1978:

> Moreover, by means of the soul or form there is a true unity corresponding to what is called the SELF [*moi*] in us; such a unity could not occur in artificial machines or in a mere mass of matter, however organized it may be.
>
> (482)

16 '[T]he physiologist and the physician must never forget that the living being comprises an organism and an individuality' (Bernard, 1865/1984, II, ii, §1, §137).
17 See, for example, Strawson 1959: 9f.
18 Furth 1978: 629.
19 See Kosman 1987 for the distinction between the realms of being, becoming and change. Cf. Dupré 2014.
20 Cf. Ackrill 1972.
21 Freeland 1987: 394.
22 Furth 1987: 39.
23 Note that the parts – in this case, the roof – also have forms (Kosman 1987: 372).
24 Aristotle *Parts of Animals* I.I 642a9–13, tr. Balme.
25 Freeland 1987: 396.
26 Wiggins 2001: 72.
27 Kosman 1987: 379.
28 Aristotle, *De Anima* II.I. 412a19–27, 412b6 (Hamlyn tr.) I will ignore, for present purposes, the considerable controversy around Aristotle's claim that *psuche* must be present in a body that has life potentially. See Wiggins 1967: 46, Ackrill 1972: 126, Williams 1978, Rosenthal 2009.
29 For example, in his most explicitly Aristotelian piece, *ISTC*, he (controversially) parses 'psuche' as 'person' (Wiggins 1967: see particularly part 4). For a critique of his interpretation, see Ackrill 1972. (Cf. Wiggins 2012 and the footnote immediately below). Wiggins has never repeated these claims.
30 What is the relation between *psuche* (form) and *phusis* (nature)? Wiggins (2012) describes the principle of activity as the *phusis*, rather than the *psuche* (which figures much more prominently in *ISTC*). From what I understand – and this is based on Gotthelf's discussion of first principles (Gotthelf 1987: 187) – the *phusis* is the general nature (e.g. the general nature say of a bird), and the *psuche* is the specific nature of the individual.

> [T]he explanation of beaks for example depends on the positing of a bird nature; this would not of course exist separately but only as a component of the natures of the individual bird-forms. Nonetheless, explanation at the level of birds would require the positing specifically of that generic aspect of these individual forms.

The general nature, or *phusis*, is the generic aspect of the specific individual form.
31 One can see the original uptake of these ideas in *ISTC*, a work partly intended as a rehabilitation of the hylomorphic doctrine.
32 See also Wiggins 2001: Chapter 3 *passim*.
33 Wiggins 2012: 8f. See also Wiggins 1995: 219, 2001: 80–81, 89.
34 Grene 1972: 42.
35 Gotthelf 1987: 222.
36 The forces at play within the environment are 'love' and 'strife' – see, for example, Schofield 2002.

37  Furth 1987.
38  Furth 1987: 44, see also Grene 1972: 403.
39  See D.M. Balme (1987: 279) for a worthwhile discussion of this anachronism.
40  There is an issue here with retrojection. It is hard to adapt Aristotle's account into modern terms, especially when talking about causality (Aristotle sees four types of causes: material, efficient, formal and final). But if nothing else, he has been (retrospectively) co-opted by the organicists, such as Woodger (see Haraway 1976: 33). It's difficult to map lines of influence but let us also recognize that Aristotle is not necessarily influential for what he actually said, rather for what he is *taken* to have said.
41  Furth 1987: 30. See also D.M. Balme 1987: 283 'The production of an animal therefore requires two material processes, which are of course combined in nature: there must be the primary actions of the elements, and there must be a limiting movement.' Here we see an upward and downward causal framework, strikingly similar to the one Dupré outlines in, for example, 2008/2012. It is Aristotelian thoughts like these which cause one to question why exactly Dupré talks Aristotle's substantialist approach to be so inimical to his own.
42  See Gotthelf 1987 and Charles 2012 for an overview.
43  See Falcon 2011.
44  Aristotle *Physics* 199a29.
45  Aristotle *Physics* 199a28.
46  Charles 2012: 255.
47  As David Charles points out, 'mysterious' is the perhaps the most prevalent criticism leveled at Aristotle's concept of final causality (Charles 2012: 228f.).
48  Unlike those who, following Marjorie Grene, hope for an Aristotelian renewal. E.g. Jonas (1966) and Kass (1999).
49  Wiggins 2001: 143 fn. 5 (and, elsewhere, 2001: 80).
50  Wiggins 2001: 143 fn. 5.
51  Correia 2008: 1013.
52  Wiggins 1967: 77.
53  Koslicki 2012: 197.
54  Aristotle *Meteorology* 390a10–13.
55  Aristotle *Metaphysics* 1040b5–10.
56  Munzer 1993: 112.
57  Aristotle *Categories* 1a–12.
58  Aristotle *De Anima* 412b2022.
59  Olson 2007: 218.
60  Olson 2007: 218.
61  Van Inwagen 1990: 15 (see also Olson 2007: 219).
62  We see similar thoughts essayed by Olson in *The Human Animal*:

> This proposal entails that an animal necessarily ceases to exist when it dies. In that case there is no such thing as a dead animal, strictly so called. We may call something lying by the side of the road a dead animal, but strictly speaking what is lying there are only the lifeless remains of an animal that no longer exists. That a dead animal should not be an animal may sound absurd. But then a ghost town is not a town, a dry lake is not a lake, a tin soldier is not a soldier.
>
> (Olson 1997: 136)

63  Wiggins 1995: 216.
64  Haack 1978: 362.

> Suppose, for instance, it should turn out that there is a type of particular, B, such that particulars of type B cannot be identified without reference to particulars of another type, A, whereas particulars of type A can be identified without reference to particulars of type B. Then it would be a general characteristic of our scheme,

that the ability to talk about B-particulars at all was dependent on the ability to talk about A-particulars but not vice-versa.

(Strawson 1959: 17f)

65 I have Alex Douglas to thank for helping me better my understanding of ontological dependence.

66 Munzer 1993.

67 Wiggins 1980: 148.

68 Hegel 1817/1975: 191–2. See also Claude Bernard's comment: 'If we decompose the living organism into its various parts, it is only for the sake of experimental analysis, not for them to be understood separately' (Bernard 1865/1984, II, ii, §1, §137).

69 There are, as far as I know, very few attempts to examine how Aristotle might respond to cases where, for example, a finger is lost and stitched back on. Stephen Munzer has written two worthwhile papers on the topic (1993, 1994).

70 The point is also helpfully drawn out by a reference Wiggins makes to J.H. Woodger, specifically in relation to biological items.

> There is one more point to be mentioned in connexion with the doctrine of the reducibility of biology to physics and chemistry: people who hold the doctrine do not in fact believe it. If you want to reduce biology to physics and chemistry, you must construct bi-conditionals which are in effect definitions of biological functors with the help of those belonging only to physics and chemistry; you must then add these to the postulates of physics and chemistry and work out their consequences. Then and only then will it be time to go into your laboratories to discover whether these consequences are upheld there. From the fact that people do *not* do this, I venture the guess that they confuse *reducibility* of biology to physics and chemistry, with *applicability* of physics and chemistry to biological objects.
>
> (Woodger 1952: 336–338)

Originally this passage featured prominently at the start of Chapter 6 in *S&S*, but has since been suppressed, and is now found in a footnote in *S&SR*, at the end of Chapter 5. Wiggins will be gratified at the attention that Woodger (oft-neglected) is now getting in the philosophy of biology (see Nicholson and Gawne 2013).

71 This is a point that Thomas Pradeu has brought to my attention.

72 Hull 1992: 184.

73 Wiggins 2001: 84.

74 Whether he *will* follow the above analysis is another question. His contribution to the collection *L'Identité Changeante de L'Individu* (edited by Edgardo Carosella, B. Saint-Sernin, P. Capelle and S.E.M. Sanchez Sorondo) suggests that he has no immediate problems with the thought that these endosymbionts are parts of us. See Wiggins 2008.

## Bibliography

Ackrill, J.L. (1972) 'Aristotle's Definitions of *Psuche*', *Proceedings of the Aristotelian Society* 73: 119–133 (reprinted in *Essays on Plato and Aristotle* (Oxford: Oxford University Press, 1997)).

Aristotle. (1936) *Physics*, W.D. Ross (ed. and trans.) (Oxford: Clarendon Press).

Aristotle. (1938) *Categories,* H.P. Cooke and H. Tredennick (ed. and trans.) (Cambridge, MA: Harvard University Press).

Aristotle. (1952) *Meteorology*, H.D.P. Lee (trans.) (Cambridge, MA: Harvard University Press).

Aristotle. (1961) *De Anima*, W.D. Ross (ed. and trans.) (Oxford: Clarendon Press, 1961).

Aristotle. (1994) *Metaphysics* (Books Z and H), D. Bostock (ed.) (Oxford: Clarendon Press).

Balme, D.M. (1987) 'Aristotle's Use of Division and Differentiae', in A. Gotthelf and J.G. Lennox (eds) *Philosophical Issues in Aristotle's Biology* (Cambridge: Cambridge University Press).

Bernard, C. (1865/1984) *Introduction à l'étude de la médecine expérimentale*, preface de F. Dagognet (Paris: Flammarion).

Charles, D. (2012) 'Teleological Causation', in C.J. Shields (ed.) *The Oxford Handbook of Aristotle* (Oxford: Oxford University Press).

Correia, F. (2008) 'Ontological Dependence', *Philosophy Compass*, 3(5): 1013–1032.

Dennett, D. (1987) *The Intentional Stance* (Cambridge, MA: MIT Press).

Dupré, J. (2008/2012) 'The Constituents of Life 1: Species, Microbes, and Genes', in J. Dupré *Processes of Life* (Oxford: Oxford University Press).

Dupré, J. (2014) 'A Process Ontology for Biology', on the *Auxiliary Hypotheses* blog, available online at http://thebjps.typepad.com/my-blog/2014/08/a-process-ontology-for-biology-john-dupré.html.

Falcon, A. (2011) 'Aristotle on Causality', in E.N. Zalta (ed.) *The Stanford Encyclopedia of Philosophy*, available online at http://plato.stanford.edu/archives/fall2011/entries/aristotle-causality/.

Freeland, C. (1987) 'Aristotle on Bodies, Matter, and Potentiality', in A. Gotthelf and J.G. Lennox (eds) *Philosophical Issues in Aristotle's Biology* (Cambridge: Cambridge University Press).

Furth, M. (1978) 'Transtemporal Stability in Aristotelian Substances', *The Journal of Philosophy* 75(11): 634–646.

Furth, M. (1987) 'Aristotle's Biological Universe: An Overview', in A. Gotthelf and J.G. Lennox (eds) *Philosophical Issues in Aristotle's Biology* (Cambridge: Cambridge University Press).

Furth, M. (1988) *Substance, Form and Psyche: an Aristotelian Metaphysics* (Cambridge: Cambridge University Press).

Ginsborg, H. (2001) 'Kant on Understanding Organisms as Natural Purposes', in E. Watkins (ed.) *Kant and the Sciences* (Oxford: Oxford University Press).

Gotthelf, A. (1987) 'Aristotle's Conception of Final Causality', in A. Gotthelf and J.G. Lennox (eds) *Philosophical Issues in Aristotle's Biology* (Cambridge: Cambridge University Press).

Grene, M. (1972) 'Aristotle and Modern Biology', *Journal of the History of Ideas* 33: 395–424.

Haack, S. (1978) 'Descriptive and Revisionary Metaphysics', *Philosophical Studies* 35: 361–371.

Haraway, D. (1976) *Crystals, Fabrics, and Fields: Metaphors of Organicism in Twentieth-Century Developmental Biology* (New York: Yale University Press).

Hegel, G.W.F. (1817/1975) *The Encyclopedia of the Philosophical Sciences*, in *Hegel's Logic: Being Part One of the Encyclopedia of the Philosophical Sciences*, William Wallace (trans.) (Oxford: Clarendon Press).

Hull, D.L. (1992) 'Individual', in E. Fox Keller and E. Lloyd (eds) *Keywords in Evolutionary Biology* (Cambridge, MA: Harvard University Press): 180–187.

Huneman, P. (2007) *Understanding Purpose: Kant and the Philosophy of Biology* (New York: University of Rochester Press).

Jonas, H. (1966) *The Phenomenon of Life: Towards a Philosophical Biology* (New York: Harper and Row/Dell).

Kant, I. (1790/1987) *Critique of Judgment*, W. Pluhar (trans.) (Indianapolis: Hackett).

Kass, L.R. (1999) *The Hungry Soul: Eating and the Perfecting of Our Nature* (Chicago: University of Chicago Press).

Koslicki, K. (2012) 'Essence, Necessity, and Explanation', in Tuomas E. Tahko (ed.) *Contemporary Aristotelian Metaphysics* (Cambridge: Cambridge University Press).

Kosman, L.A. (1987) 'Animals and Other Beings in Aristotle', in A. Gotthelf and J.G. Lennox (eds) *Philosophical Issues in Aristotle's Biology* (Cambridge: Cambridge University Press).

Leibniz, G. (1695/1978) *Système Nouveau de la nature et de la communication des substances aussi bien que de l'union de l'âme avec le corps*, in G. Leibniz *Die Philosophischen Schriften*, G.J. Gerhardt (ed.) (vol. 4, reprint, Hidesheim: Georg Olms): 471–504.

Munzer, S. (1993) 'Aristotle's Biology and the Transplantation of Organs', *Journal of the History of Biology* 26(1): 109–129.

Munzer, S. (1994) 'Transplantation, Chemical Inheritance, and the Identity of Organs', *The British Journal for the Philosophy of Science* 45(2): 555–570.

Nicholson, D. and Gawne, R. (2013) 'Rethinking Woodger's Legacy in the Philosophy of Biology' *Journal of the History of Biology* (Springer).

Olson, E. (1997) *The Human Animal: Personal Identity Without Psychology* (Oxford: Oxford University Press).

Olson, E. (2007) *What Are We? A Study in Personal Ontology* (Oxford: Oxford University Press).

Rosenthal, D. (2009) 'Aristotle's Hylomorphism', *Philpapers,* available online at http://philpapers.org/rec/ROSAH.

Schofield, M. (2002) '*Empedocles*', in E. Craig (ed.) *Routledge Encyclopedia of Philosophy* (London: Routledge). Available online at www.rep.routledge.com/article/A046SECT5.

Strawson, P.F. (1959) *Individuals: An Essay in Descriptive Metaphysics* (London: Methuen).

van Inwagen, P. (1990) *Material Beings* (Ithaca: Cornell University Press).

Wiggins, D. (1967) *Identity and Spatio-Temporal Continuity* (Oxford: Blackwell).

Wiggins, D. (1980) *Sameness and Substance* (Cambridge, MA: Harvard University Press).

Wiggins, D. (1995) 'Substance', in A.C. Grayling (ed.) *Philosophy 1: A Guide Through the Subject* (Oxford: Oxford University Press.

Wiggins, D. (2001) *Sameness and Substance Renewed* (Cambridge: Cambridge University Press).

Wiggins, D. (2008) 'Deux Conceptions d'Identité', in E. Carosella, B. Saint-Sernin, P. Capelle and S.E.M. Sanchez Sorondo (eds) *L'Identité Changeante de L'Individu* (Paris: L'Harmattan).

Wiggins, D. (2012) 'Identity, Individuation and Substance', *European Journal of Philosophy* 20(1): 1–25.

Wiggins, D. (2016) *Twelve Essays* (Oxford: Oxford University Press).

Williams, B. (1978) 'Hylomorphism', Princeton Classical Philosophy Colloquium, 7 December, 1978.

Wolfe, C.T. (2010) 'Do Organisms have an Ontological Status', *History and Philosophy of the Life Sciences* 32(2): 195–231.

Woodger, J.H. (1952) *Biology and Language* (Cambridge: Cambridge University Press).

# 9   Brain transplantation

Death interests and worries us in equal measure. It may not always be at the forefront of our minds but few among us go through life without thinking of its end. We are concerned about our survival; we are concerned about those situations that we cannot survive. On a very intimate level, we are interested in human persistence conditions. Nor are our concerns simply about biological functions, about whether our hearts continue to beat or our blood continues to course. We look to memory-loss and dementia as well, and wonder whether these are the sorts of things we can suffer and stay the *same*. These are real concerns, which can achieve a strange and frightening prominence in our thoughts, and merit – if anything does – serious philosophical attention.

Moreover, while human life expectancy has certainly increased as a result of developments in healthcare, we are not always sure that these technologies – such as 'life-support' machines – really *do* safeguard our survival. Would you, do you think, continue as a vegetative state patient lacking higher cortical functions? How many side-effects of extensive drug therapies could you yourself endure (mood-swings, amnesia, psychotic episodes, hormonal imbalances)? These questions sustain the personal identity debate as it appears in Anglophone philosophy; and while our interest in the metaphysical relation of 'identity' may range, it is these questions which are closest to our hearts.

To sharpen our focus on such issues, philosophers sometimes discuss cases that straddle the divide between science fact and science fantasy. There are stories about cloning (an extant technology – though strictly regulated), which encourage us to inquire deeper into the notion of human sameness. Similarly – and at the fantastical end of the spectrum – we have tales about teleportation, fissioning and cryogenic stasis. Are these procedures you would submit to? Do you think the individual that emerges from the teleportation chamber would be *identical* with you?

Sydney Shoemaker's canonical 'brain transplantation' narrative is a staple of the 'personal identity' literature and the standard measure by which accounts of human individuation are measured. Wiggins returns to it time and again; yet by my lights Brown's peculiar adventures seem somehow to resist his analysis. The aim of this chapter is to investigate why it does so, whether he can give a principled response to the story, and whether this is a response that we would want to echo.

To begin with, I set out Wiggins's various critical treatments of the story. Then I re-examine Shoemaker's text. I argue that the idea of brain transplantation subtly shifts our metaphysical focus away from an *organism* onto a biological item whose parts are *ontologically independent* of the whole. Shoemaker's scenario is in a relevant sense *mechanistic*. This reading is supported by a historical review, in the Appendix, of the origins of the story in Locke. (I hold that the Lockean discussion of personal identity is a response to tensions between the doctrine of the resurrection and corpuscularian forms of biological mechanism.)

Can an *organism* suffer the sort of disassembly that brain transplantation involves? Drawing upon Chapter 8, I shall argue that Wiggins has stronger grounds than he recognizes to deny that it can. This claim will provoke objections. More anodyne cases of transplantation – such as heart transplantation – involve similar kinds of disassembly. Does the organism survive heart transplantation? And if not, does the *person*? These questions take me back to the worries that I had with the Human Being Theory in Chapter 5 and provoke a recommendation for Wiggins.

## 1  Changing perspectives

Sydney Shoemaker's story about the unfortunate Brownson appeared in 1963 in *Self-Knowledge and Self-Identity*. The text, which describes the transplanting of Brown's brain into Robinson's body, is quoted and analysed below. The aim of this section, meanwhile, is to present an overview of Wiggins's response to this peculiar philosophical fairy-tale. It will mainly refer to eight of Wiggins's texts, spanning his philosophical career: *Identity and Spatio-Temporal Continuity* (1967) 'Locke, Butler and the Stream of Consciousness' (1976), *Sameness and Substance* (1980), 'The Person as Object of Science' (1987), 'Reply to Snowdon' (1996), *Sameness and Substance Renewed* (2001), 'Reply to Shoemaker' (2004) and 'Identity, Individuation and Substance' (2012 – now revised for his forthcoming collection, *Twelve Essays*).

In *ISTC* and in the context of his treatment of the logic of individuation, Wiggins arrived in Part 4 at the question of personal identity. At that time, there was widespread controversy between upholders of a 'bodily' criterion and a 'memory' criterion. He was eager to show that neither position was satisfactory ('no correct spatio-temporal criterion of personal identity can conflict with any correct memory-criterion or character-continuity criterion of personal identity').[1] That text constituted a search for a better account of the material continuity of a person than a simple bodily account. This was to be a *functional* account, which made room for an individuative nucleus that was the brain, seen as the core of the central nervous system. It was the logical need for a better criterion of personal continuity that explained the force of Shoemaker's thought-experiment, and Wiggins's regard for it in *ISTC*.

> The kind of individual we are to define is not made of anything other than flesh and bones, but, unlike the body which it at some times shares its

matter, it has a characterization in functional terms which confers the rôle, as it were, of *individuating nucleus* on a particular brain.[2]

The crucial point is that if Brown survives it is because his *life* goes along with his brain:[3]

> [W]hat matters ... is the continuity of Brown's life and vital functions as they are planted in one body and recognizably and traceably transposed in another body.[4,5]

However, even in *ISTC*, Wiggins is struggling with the problem – which he sees as a problem for the logic of individuation – that the brain is symmetrical and its two halves are (almost) equipollent.[6] If transplantation *were* possible, the two hemispheres could be separated and transplanted into two separate husks – resulting in splinters of the original Brown: Brownson (1) and Brownson (2). Wiggins's adherence to Leibnizian identity means he cannot read both splinters to be identical with Brown. Nor does he see a principled reason for privileging one over the other. So Brown cannot, on pain of logical transgression, survive fission.[7] And since there is no relevant difference between the procedure that produces Brownson and the one that produces the splinters, we will – he suggests – be encouraged to think that Brown *does not* survive brain transplantation. Wiggins's response in *ISTC* is to incorporate a 'one parcel' criterion into his account:

> Coincidence under the concept *person* [requires] the continuance in one organized parcel of all that was causally sufficient and causally necessary to the continuance of essential and characteristic functions, no autonomously sufficient part achieving autonomous and functionally separate existence.
>
> [But] irreducibly psychological concepts are required to define an entity with the right principle of individuation to be a person.[8]

In his later work, this one parcel stipulation is seen to be an unstable resting place (though, in light of the reading below, it may regain some of its former appeal).[9] He claims (in *S&SR*) that more is required to show *why* this condition inevitably and naturally arises from the proper understanding of *person* and *human being*.[10] Without further philosophical analysis the stipulation seems arbitrary.

Wiggins's reading of brain transplantation in 1967 is a tentative one. If Brown survives it is because the brain contains – in some way – everything that is essential to the person (though the person is not the brain). Yet this analysis is saddled with doubts. In addition to being obscure about how the brain 'contains' or 'seats' or 'houses' an individual's life,[11] there is the worry concerning fission.

It is very likely because of these worries that Shoemaker's story largely disappears from Wiggins's work over the following decade. Indeed, in his 1976 paper, and in *S&S*, the main additions to Wiggins's thoughts about brain transplantation are two further doubts. In an almost incidental footnote in 'Locke, Butler and the Stream of Consciousness', he criticizes Shoemaker's narrative along the following lines (and in this he follows Bernard Williams' thoughts in the celebrated paper, 'Personal Identity and Individuation'):[12]

> How do we fix the brain to the physiognomy of the new body which is to receive it?... How is the existing character expressed in the new body? We are deceived by the quality of the actors and mimics we see on the stage if with the help of greasepaint and props they have made us think this is as (relatively) simple as the transposition of music from one instrument to another.[13]

This point about the relevance of physiognomy is reiterated in *S&S*[14] along with a general complaint about the comprehensibility of brain transplantation:

> [W]e should take nothing for granted about how well we really understand brain transfers of the kind described by Shoemaker.[15]

In both the 1976 and 1980 texts Wiggins refrains from engaging with the Brown/Brownson story. As he sees it, the case has yet to be made for the possibility of this kind of transplantation. And while he discusses fission (which gave rise to the earlier worries) he does not do so in relation to hemispherectomies and transplants, but rather Lamarckian inheritance and branching.[16] His problems with brain transplantation fall into the background to be replaced by his concerns with 'quasi-memory' (discussed above in Chapter 4). Derek Parfit had argued – in 'Personal Identity' (1971) – that the right moral to draw from Wiggins's chapter in *ISTC* is that there could be survival without identity. Doing so, he invoked Shoemaker's *q*-memory – and it was the frailty of that notion which received Wiggins's attention in the articles from that period.[17] In 'The Person as Object of Science' (1987) there is no explicit mention of brain transplantation or Shoemaker, despite his review there of past works, and his discussion of neo-Lockeanism.

It is Paul Snowdon's 'Persons and Personal Identity' that encourages Wiggins, in his 'Reply' (1996), to re-engage with Shoemaker's story. And following the development of his 'Animal Attribute View' in his 1980 and 1987 texts, Wiggins has a new resource to draw on when dealing with the narrative. He reiterates past doubts, but writes that *if* the neo-Lockeans are right that the person goes with the brain, since the concept *person* is concordant with the *human being* concept, the human being will necessarily go as well:

> [I]f I *must* allow survival, I am not sure why I am committed to denying that the survivor who emerges from all these goings on is the same *human being* or the same *animal* as the one who entered them.... I have insisted on the dependence of the concept person upon the concept human being.... [O]nce you understand what a *human being* is and what the *seat of consciousness* is,

surely you will not too readily assume that you know what it would be for a human being to be given a new seat of consciousness. If transplantation really were possible, then would not the person follow the seat of consciousness? In that case, does not the animal that the survivor is follow it too?[18]

It is not claimed that Brown would be identical with Brownson. Rather, Wiggins is saying that *if* such an operation were possible one might think that the animal, like the person, would go along with the brain. Given his interlocutor, the passage is best read as a critique of Animalism. Snowdon and his allies have not escaped any of the problems that confront the neo-Lockeans (with whom Wiggins has made his very conditional peace). If the animal survives brain transplantation, the animalists must reckon with the same issues with fission.

Contained within these texts we find various reasons for rejecting Shoemaker's story. Prime among them are Wiggins's concerns with the relevance of physiognomic differences and his doubts about fission. In *S&SR*, the first extended treatment of transplantation since 1967, he specifically addresses these issues. In its final chapter, he re-emphasizes Williams's worry with physiognomy, writing of his earlier self:

> I ought to have been much more troubled by the point that, even if Brownson could talk like Brown ... he could scarcely have stood and walked and run and jumped and smiled and sulked and earnestly entreated and frowned and laughed like Brown.[19]

It is a forceful point, but I am not sure how far it will take him. In discussion with Shoemaker in *The Monist* (2004), the possibility of brain transplantation between identical *twins* is raised (this is latterly supplemented by Shoemaker's talk of transplantation between clones).[20] Wiggins sees this shift to be a concession ('I note in passing that it would have been gracious for Shoemaker to remark on the magnitude of this concession'),[21] but I am unsure as to why. There will, of course, be physiognomic differences between monozygotic twins. But the same is true, for example, of an individual before and after a car accident, and we do not doubt that organisms can survive such calamities.

What then of the fission objection? In *S&SR*, Wiggins re-emphasizes the need for a principled reason to distinguish the procedures that resulted in Brownson and in Brownson (1) and (2).[22] Wiggins attempts (on Shoemaker's behalf) to provide such a reason based on the Butlerian claim that memory presupposes and thus cannot constitute identity. Taking Butler's line would allow one to say that in fission cases the splinters, while seemingly remembering Brown's life, cannot actually be said to do so. The memories presuppose identity and, since the splinters are not identical with Brown, theirs cannot be real memories. Because Brownson genuinely remembers Brown's life, and the fissioning procedure that produced the splinters prevents them from doing so, the two stories may be said to be relevantly different. If this argument holds then there is no reason, on the grounds of fission, to disallow brain transplantation.

However, having regulated these objections, Wiggins immediately raises two further doubts in *S&SR*. The first was alluded to in 1980 and in 1996 and concerns the *conceivability* of these kinds of narratives. In Shoemaker's original text, there is no real examination of the kinds of revolutionary technological advancements that would be required by the procedure. Thus, Wiggins points out:

> [P]hilosophers are still apt to underestimate the preternatural dexterity and knowledge that the imaginary surgeon and [their] equally imaginary team of anaesthetists, suturists, radiographers, laser-technicians, physiotherapists, psychotherapists, counsellors and the rest, would have to bring to bear.[23]

This is a fair point but not a critical one. It cannot rule out Brown's survival as Brownson. It is followed, however, by a more specific concern which relates to the notion of a *guarantee*. Brain transplantation might seem, Wiggins suggests, to violate the lawful dependability of the natural biological process, and thus individuals like us cannot be said to survive such procedures:

> A genuine guarantee relating to this or that process must relate to the nature of the process itself rather than a mere description of it. Moreover, genuine guarantees exist. However you describe it, the process of jam-making can be guaranteed not to produce heavy-water out of ordinary water.... Another process that comes with a certain guarantee is the natural process, sustained by the operation of numerous laws of biochemistry, physiology and the rest, by which a human being comes into existence and matures, and eventually ceases to be, by 'natural death'. That process is *not* of course guaranteed to save a human being from murder or from premature death by asbestosis, say, or irradiation. But it is certainly guaranteed not to produce multiples, not to transplant brains or half-brains, and not (if that were the better way to think of Brownson) to furnish new bodies to living, continuing brains. That is what makes this familiar process and the principle associated with it one part of the basis for the making of judgments of identity.[24]

It is on the basis of the *dependability* of the process that we can distinguish between judgments of identity, and judgments 'to the effect that the object *b* is the proper replacement/surrogate/proxy for object *a*'. The suggestion is that brain transplantation violates this guarantee and that the living activity is thereby undermined.

Will this allow him to reject Shoemaker's story? Wiggins himself points out that violations of the guarantee do not always stop the process or imperil identity. For example, there are those small procedures – orthopaedic, osteopathic, etc. – in which 'the substance's organic independence can still be conceived as undiminished'.[25] At the other end of the spectrum, interferences like teleportation violate the guarantee to such an extent that we lose track of the original process. In such instances 'we have lost hold altogether of the notions we began with of what Brown is'.[26] The natural substance has become 'artefact-like', something not so much to be encountered in the world as putatively made or produced by us, something that it is really up to us (individually or collectively) not merely to heal or care for or protect but also to repair, to reshape, to reconstruct ... even to reconceive.[27]

The thing that is so unsettling about the surgeon's experiment as I see it is that it spans the divide between, on the one hand, natural substances (which have their own inherently orderly ways of enforcing some individuative decisions ... while forbidding others), and, on the other hand, artifacts, where individuative thought is forced into an opportunism that it has to strive constantly (and retrospectively sometimes) to make principled.[28]

This 'denaturing' of the natural substance is the topic of the sections that follow.[29] The point here is that Wiggins raises a question which he leaves unanswered: *why* is brain transplantation a technological interference that organisms, human beings, cannot suffer? *If* we think of Brownson as Brown, then the conceptual consilience between the *person* concept and the *human being* concept will have been undermined – and *human person* will converge more closely on the conception of an artefact. But again, this is an important 'if' which brooks any number of 'buts'.

Since giving his semi-positive response in 1967, Wiggins has repeatedly iterated worries and doubts about Shoemaker's story. As demonstrated, *S&SR* delivers a bumper crop. And in 2004, in a fiery exchange with Shoemaker in *The Monist*, we find perhaps the most explicit denial of the claim that Brown survives as Brownson:

> If Brownson isn't Brown, then something peculiar must have happened. But it has. There has happened the intervention, as I am now imagining the case, of a mad surgeon. What shall we say the surgeon has done? ... Let us say that out of rather unpromising natural materials – for Brownson has little or no resemblance to Brown – the surgeon has tried to make a working model of Brown. ... He has created a living something-or-other which, in addition to its other sufferings, is a repository or receptacle for mental events that are downwind of an ordinary human life.[30]

How persuasive is this ruling? It is supported by the physiognomy objection which, it has been suggested, loses its force when faced with identical twins. There is no talk of fission cases or conceivability or of guarantees. Has he rescinded these later objections? Can the earlier ones be somehow bolstered? It is not clear. His latest analysis of the case, in his Mark Sacks lecture in 2012, is much the same. He reasserts his worry about guarantees and physiognomy,[31] but his final considered opinion remains obscure:

> In discussion of Brown and Brownson it is often assumed that, if there were later mental events that were downwind from Brown before brain surgery,

then the person Brown – or some person Brown – must have persisted somehow. But how strongly can a well-founded conception of person support that assumption? Again I am unsure.[32]

Despite prolonged discussion there remains something unsatisfying about Wiggins's analyses of Shoemaker's story.[33] My aim in the next section is to pull together strands from the previous chapters and doing so to present a principled reason, not only for denying the identity of Brownson and Brown, but for reassessing Shoemaker's narrative more generally.

I have claimed that, for Wiggins, our method of partitioning the biological realm depends on us being the kinds of being that we are. We pick out 'organisms'. In Chapter 8 it was argued that we understand organic substances as *genuine unities*. When it comes to our everyday navigation of the world, we are *anti-reductionist*. In the remainder of this chapter, I will claim that Shoemaker's brain transplantation story implicitly shifts the metaphysical focus from *organisms* onto another type of entity: in virtue of an underlying mechanism it directs our attention onto the kind of being that can be *disassembled and reassembled*. Imagining the case, we pick out *something* – but not a biological being as ordinarily conceived (and here, 'ordinarily' has the full force of Wiggins's descriptivism). Following Wiggins's comments in *The Monist*, we may say, for the time being, that our focus has been shifted onto a 'living something-or-other'.[34]

Let us start with some general methodological issues with thought experiments before examining how Shoemaker's story implicitly, if unknowingly, smuggles certain conceptual assumptions into the personal identity debate.

## 2 Methodological concerns

Science-fiction thought experiments like the brain transplantation case are used to draw out intuitions and to investigate what we *mean* exactly by certain terms. By those who deploy them, they are seen to be relatively innocent philosophical tools. However, a growing number of theorists – among them, Michèle le Doeuff, Susan James and Margaret la Caze[35] – have subjected these narratives to closer scrutiny and argue that, rather than simply being illustrative or pedagogically useful devices, they often play significant roles in the argument within which they are deployed.[36] They are, says la Caze: 'extremely important to the expression of philosophical thought, to the way debates are structured, and assumptions are shared; they can also work to persuade and provide support for a particular view and exclude alternative views and methods.'[37] This should be borne in mind when considering the tale of Brown, Robinson and the unfortunate Brownson (quoted here in the abridged form in which it appears in *S&SR*):

Suppose that medical science developed a technique whereby a surgeon can completely remove a person's brain from his head, examine or operate on it, and then put it back in his skull (regrafting the nerves, blood-vessels, and so forth) without causing death or permanent injury.... One day a surgeon discovers that an assistant has made a horrible mistake. Two men, a Mr Brown and a Mr Robinson, had been operated on for brain tumours, and brain extractions had been performed on both of them. At the end of the operations, however, the assistant inadvertently put Brown's brain in Robinson's head, and Robinson's brain in Brown's head. One of these men immediately dies, but the other, the one with Robinson's body and Brown's brain, eventually regains consciousness. Let us call the latter 'Brownson'.... He recognizes Brown's wife and family (whom Robinson had never met), and is able to describe in detail events in Brown's life, always describing them as events in his own life. Of Robinson's past life he evidences no knowledge at all. Over a period of time he is observed to display all of the personality traits, mannerisms, interests, likes and dislikes, and so on that had previously characterized Brown, and to act and talk in ways completely alien to the old Robinson.

What would we say if such a thing happened? There is little question that many of us would be inclined, and rather strongly inclined, to say that while Brownson has Robinson's body he is actually Brown. But if we did say this we certainly would not be using bodily identity as our criterion of personal identity. To be sure, we are supposing Brownson to have *part* of Brown's body, namely his brain. But it would be absurd to suggest that brain identity is our criterion of personal identity.[38]

Since its debut in Shoemaker's *Self-Knowledge and Self-Identity*, this narrative has been the subject of considerable discussion, of numerous refinements and varied critiques.[39] Prime among the latter is a strand of feminist criticism, which questions the binary distinction that underpins the story. In her brilliant study, 'Feminism in philosophy of mind: The question of personal identity',[40] Susan James shows how such 'character transplants' implicitly enforce a worryingly thin notion of, all but disembodied, personhood.

[T]he body is thought of as a container or receptacle for character. The brain figures as a container in which a person's psychological states can be preserved, and the body figures as a more elaborate receptacle for the brain.

Properties which do not fit neatly into the category of the psychological are held to be marginal or irrelevant to character.[41]

Shoemaker's story reiterates the traditional division between the psychological and the bodily. Firstly, we see the marginalization of the *body* (historically, symbolically feminine) and the privileging of the *mind* (historically, symbolically masculine).[42] Secondly, and simultaneously, we see how the anonymising of the body affirms a particular view of character; expressly *embodied* character traits – like one's dexterity or sexuality – are held to be irrelevant.[43] Shoemaker's story misses out what we may well see to be central aspects of our personhood, our experience of ourselves as distinctively *embodied* beings. These facts are fundamental to our being the kinds of entities that we are and should thus figure in our understanding of human persistence. (And this last point – as will be clear from the discussion above – is one with which Wiggins is in total accord.)[44]

These kinds of criticisms have resulted in refinements to the story. Now, rather than construing the donor and recipient's physiology as marginal, philosophers (including Shoemaker) specify that the transplantation occurs between monozygotic twins (or clones).[45] Yet one may think that worries remain, and the aim here is to expose – in a similar vein – another suppressed assumption contained within the narrative.

## 3 A suppressed assumption

Consider the story again. Better yet, consider the standard procedure – 'brain extraction' – the corruption of which produces 'Brownson'.

> [It is] a technique whereby a surgeon can completely remove a person's brain from his head, examine or operate on it, and then put it back in his skull (regrafting the nerves, blood-vessels, and so forth) without causing death or permanent injury.[46]

The patient described here is the kind of thing that can be cut into pieces and then reassembled, *and not die* (or indeed suffer permanent injury). There is nothing but technical dexterity to stop Brown being split into still more pieces. The heart, the lungs, the kidneys, the stomach might be separated – and so long as they are reassembled, the story goes, the organism survives suffering neither 'death or permanent injury'. The work above suggests this aspect of the thought experiment merits closer attention.

The first thing to note is that there is something interestingly reminiscent here of Wiggins's discussion of clocks, as artefacts, mentioned in Chapter 3. He writes that 'the repair of a clock ... permits both disassembly and replacement of parts. We do not look back to the time when a clock was being repaired and say that the clock's existence was interrupted while it was in a dismantled condition.'[47] Clocks, and other artefactual entities, are the kinds of things that can undergo disassembly and reassembly (this is what provoked all those puzzles about identity); and in Shoemaker's story the human body appears to exhibit this self-same quality.

The second thing to point out is that this association between living bodies and artefacts, or 'mechanisms', is in good – or at least long – standing. In Shoemaker's story we are invited to think of the patient as though he (Brown) were *a human-engineered machine*. Thus the transplantation narrative corresponds to what was briefly described, in Chapter 7, as the 'mechanistic' analysis of biological beings.[48] Brown is being seen mechanistically, as realising some of the distinctive features of certain mechanical artefacts.

These considerations lead us towards the thought that transplantation, of the kind described by Shoemaker, encourages a form of 'biological mechanism'. This suggestion is not a novel one; we find it, for example, in Ian Hacking's extensive discussions of transplant medicine. In his paper, 'Our Neo-Cartesian Bodies in Parts',[49] Hacking argues that the technological advancements that allow us to separate and reintegrate organs tangibly reinforce this 'mechanistic' conception of the biological world. To see ourselves as the potential subjects of transplantation is to see ourselves as in some way *like machines*.

> It is seldom noticed that we seem to be edging closer to fulfilling a simplistic version of a Cartesian dream, whereby bodies are just machines in space, composed of machine parts.... Something like Descartes's two categories may be forced on us again as the result of our technological prowess.[50]
>     The bodily revolution may be a revolution in that sense – the reinstatement of a Cartesian attitude to the body as a machine. It has become a machine, subject to engineering projects large and small.[51]

There are some powerful insights here – about how technology affects our everyday experience of the world – and we will return to these. For the time being though, let us continue to examine this conceptual association, which Hacking finds in transplantation surgery, between mechanisms and living beings. Let us clarify 'mechanism'. Like 'reductionism', it captures a variety of inter-related theses which collect around the powerful guiding metaphor of the *machine*. It has, as discussed, a methodological aspect. More important, for present purposes, is its metaphysical aspect: it is powerfully linked to the causal reductionist's thesis. Mechanisms – clocks and cars and so on – exemplify causal reductionism. A mechanism, as a whole, exerts no relevant, novel causal influence over its parts. There is no emergent causal property at the higher level. A clock (say) has its causal powers in virtue of the causal powers of its parts. This feature of mechanisms is widely commented on in this connection by, among others, Dupré (in his Spinoza lectures):

> A good machine starts with all its parts precisely constructed to interact together in the way that they will generate its intended functions. The technical manual for my car specifies exactly the ideal state of every single component ... failing components can be replaced with replicas, close to the ideal types specified in the manual.... Reductionism is almost precisely true of a car. We know exactly what its constituents are – they are listed in the manual – and we know how they interact: we designed them to interact that way.[52]

In transplantation this metaphysical characterization of mechanisms is transposed onto the human body. And in Shoemaker's narrative, a successful 'brain extraction' causes neither death nor permanent injury; the patient is assumed to 'live through' the procedure. Peering into the operating theatre an imagined bystander might plausibly say that the 'dismantled' patient is still alive (indeed, that they are undergoing the operation)[53] – and understanding the patient thus is to see her or him as an entity that can survive disassembly in the same way as a mechanism can. Extending this thought, one might claim that the whole (the patient) does not exert any relevant causal influence over its parts because it persists while these parts are separated. The reintegration of the parts may make things *quantitatively* more complex, but there will be no relevant *qualitative* difference.[54] As Hacking points out, this kind of disassemblable beast is of the same family as Descartes' bête-machine (a point reinforced below, in the Appendix). It is the kind of entity described in a scientific framework which focuses on parts and their parts, and the parts of those parts – and less on the unified whole.

The logic of transplantation is reductionist. Still, the successes of transplant medicine do not necessarily underwrite the truth of the causal reductionist's thesis. There is clearly a point where the metaphor breaks down. A mechanism's components – the springs of a clock, for instance – do not, when separated, deteriorate in the same way as do the bodily organs. A heart that has been extracted for transplantation does not survive for very long (even put on ice). A spring, by contrast, deteriorates in no qualitatively different a fashion to when it is integrated into a clock. And unless Shoemaker is happy to abandon even the veneer of plausibility, his surgeon will not be able to simply place the patient's brain on a sideboard to await reinsertion. It must be transfused with blood or frozen (etc.).

Given this, it remains open whether or not the biological whole exerts a relevant causal influence over its parts. It clearly sustains them. Perhaps it does so in virtue of a non-derivative causal power. Perhaps this power can be simulated (during brain extraction) by the doctors performing the procedure. The fact that medics can transplant organs does not commit one outright to the thought that novel causal powers only exist at the lower levels.[55] So an emergentist reading of transplantation remains possible. (Indeed, Dupré's own processualism seems to accommodate the story quite neatly; it is a distinctive feature of processes that they can have spatially discontinuous parts. A process – the life of a bee colony, say – need not occur in some spatially contiguous region.) The point here, however, is that whatever the entity in Shoemaker's story is, be it the reductionist's bête-machine or Dupré's spatially dispersed process, it is not an *organism*, in so far as an organism is a *natural substance*.

## 4 Organic unity (again)

I have argued that the individuation of human beings – or human 'organisms' – involves individuating them as genuine unities. We would fail to function in the way that we do, in our characteristically human fashion, if we thought that animals were nothing more than the sum of their parts (or processual beings). We might also say, remembering the comments offered in Chapters 5 and 7, that Wiggins sees the notion of a unified being to be central to our moral interests and practices; you are concerned for *me*, I am concerned for *you*; we are beings with individual biographies, 'passions, thoughts and actions',[56] separate and discrete (though, of course, part of broader society).[57] These thoughts have been teased out of Wiggins's texts – there is no explicit, sustained statement of them in his metaphysical works – yet attending to them we can see how inimical the brain transplantation story is to his metaphysics of organic substances.

Unity was explained in relation to *ontological dependence* and 'mechanism' can be seen to capture a claim, expressible in those terms, about the metaphysical relation between the parts of the body and the whole. I contended that, in genuine unities, the parts ontologically depend on the whole. Significantly, this view of part – whole dependence contrasts with the way we understand the relation between a mechanism and its components. Thinking of a spring does not involve thinking of a clock. Thinking of a screw does not involve thinking of a catapult or a computer (etc.). Screws and springs are parts that function in numerous, varying contexts – thus a statement of the essence of a screw does not have to make reference to a mechanism of which it is a part.[58] The Aristotelian distinction between different part-whole dependence relations, is helpfully rendered by Robert Pasnau (the following quotation comes from his 'Form, Substance, and Mechanism'):

> [I]n the case of genuine substances, the parts are radically dependent on the substance for their continued existence. Take away a piece of flesh and it becomes something else. This is not the same for non-substances. Take a brick away from a house, and it remains a brick. So the substance is unified not because it can exist without its parts, but because its parts cannot exist apart from it.[59]

One notable feature of biological mechanism, as expressed in the transplantation story, is that it dissolves this distinction. For the mechanist, biological parts are understood to stand in the same dependence relation to biological wholes as a mechanism's components stand to the mechanism. Biological parts can be seen to exist even when separated from the organized whole. Consequently, biological items are conceived of – like mechanisms – as the kinds of things that can suffer disassembly.

In accepting that humans could, in principle, undergo the sort of dismantling described by Shoemaker, one is committed to this mechanistic thesis (if not the causal reductionist's thesis – though the links between these are addressed in the Appendix). In the transplantation narrative, the persistence of the brain does not depend on it being integrated into a biological unity and this is tangibly at odds with the 'organismic' Aristotelian picture that was recommended to Wiggins in Chapter 8. If Wiggins could be convinced of the claims I have made about the *unity* of organic substances then he would be in a position to a give a more forceful and direct critique of the position that he opposes. Shoemaker's story *smuggles in* a particular and problematic view of biological entities.[60]

## 5  A shift in metaphysical focus: 'human persons as artefacts?'

In Shoemaker's thought experiment, Brown is conceived of as a living thing that can be *dismantled* – conceived that is as endowed with a metaphysical character quite other than that of an unified/unitary organism (in the neo-Aristotelian sense). Shoemaker's story thus subtly shifts our attention from one metaphysical entity onto another, and so our thoughts (and Wiggins's) about what persists are confounded. Helpfully, in *S&SR*, Wiggins describes a comparable shift in metaphysical focus:

> [After fission], one very easily falls into thinking of Brown as a thing that persists in Brownson (1) *and* Brownson (2). One conceives of Brown as a thing that persists in its/his instantiations, a thing that is wherever they are – just as the sail that lies over you, over me, and over a friend of ours, is where I am, where you are, *and* where he is…. In short, one thinks of Brown as a concrete universal. Nothing need be wrong with that…. Nevertheless, Brown reconceived as such a concrete universal is not the sort of thing whose survival was to have been described in the case of Brownsons (1) and (2). Nor is this how we conceive of subjects of experience when posing the questions of personal identity.[61]

As a result of technological interference our metaphysical focus drifts from the *substance* that we take Brown to be (and which we typically refer to when judging persistence) and onto a *concrete universal*. What is a concrete universal exactly? It is, roughly, the collection of material instantiations of a specific type, a group of clones, or copies, or reproductions.[62] In the introduction to his *Twelve Essays*, Wiggins gives the example of a 'genotype' as a concrete universal, of which various propagules are instances[63] – and we find a refinement of this idea in the revised version of his 2012 paper found in that collection:

> Suppose that we confront at a given moment an amoeba, a thing which perpetuates its lineage by dividing. The amoeba we confront cannot be traced through its divisions as one and the same substance. The lineage is a particular lineage but scarcely a particular substance. The lineage is something realized and instantiated by short-lived particulars. It is a sort of universal.[64]

There has been a growing focus on 'concrete universals' in Wiggins's work and the extension of the passage in this revised paper is one example. In *S&S*, the claim that Brownson (1) (or (2)) was a part of a concrete universal was closer to a *reductio ad absurdum*, intended to demonstrate the absurdity of these claims about survival; more recently, he has begun to take these plural entities more seriously (motivated by concerns with other entities – like clonal plants and amoebas). Either way, the guiding thought remains the same: concrete universals have a different metaphysical character from substances.

> Unlike a universal/type/sort/kind/clone/character, a substance does not have specimens or instances. Nothing falls under it, exemplifies it or instantiates it.[65]

A concrete universal has spatially discontinuous parts (entities like the Brownson splinters or the members of the lineage). It is not genuinely unified in the way that a substance is; it does not realise a principle of activity in the same way. It is not necessary to go into all the differences. The important point is that, in the case of the splinters, the shift is not a shift in our conception of the substance, Brown: we are not conceiving of the same thing in a new way. Rather, we have latched onto an entity of a different sort[66] (which may, at a moment, inhabit exactly the same space as the substance). My proposal is that the brain transplantation story should be read along similar lines. It jogs our focus; our attention moves from a *substance* onto a 'living something-or-other' which has parts that are *ontologically independent* of the whole. Can we give this thing a name, one that is better than Wiggins's gestural attempt? I think we might usefully call it a *biological artefact*.

Consider again the short passage, from *S&SR*,[67] where Wiggins discusses the effects of massive transplanting of organs.

> [T]he conception of a human person will diverge further and further from that of a self-moving, animate living being exercising its capacity to determine, within a framework not of its own choosing and replete with meanings that are larger than it is, its own direct and indirect ends. The conception will converge more and more closely upon the conception of something like an *artefact* – of something not so much to be encountered in the world as putatively made or produced by us, something that it is really up to us (individually or collectively) not merely to heal or care for or protect but also to repair, to reshape, to reconstruct … even to reconceive.[68]

In this passage, he is suggesting that transplantation leads us to think of persons – and thus human beings – *artefactually*. And, while there are various readings of his view of artefacts, he might say here, in line with the thoughts advanced in Chapter 3, that they have a determinately different ontological profile from substances. At points he appears to think that they are definitionally unable to exemplify the central elements of our pre-theoretical *substance* concept – but I suggest that the fecundity of his metaphysics allows him to go beyond this negative verdict.

Working within the descriptivist tradition, Wiggins turns to the structure of our thoughts to guide his metaphysics. It is not unimportant, therefore, that we interact with, and think about, artefacts in unique and distinctive ways.[69] We treat them in a different manner from the manner in which we treat living things (and it is around the living things that the *substance* concept crystallizes).[70] Considering the ever-expanding literature on this subject, we might venture to say that *artefact* (or *tool*), alongside *person* and *organism*, is another central category of our pre-theoretical framework.[71] Neuroscientists are fascinated by the neurological structures that appear to facilitate our thoughts about tool-use[72] and, in the sphere of behavioural and brain sciences, artefact-use and manufacture is routinely positioned as a marker of cognitive discontinuity: creatures that can make and use tools think about the world differently from those that cannot.[73]

Nor is it an accident that there is such emphasis in the phenomenological tradition on our pre-theoretical engagement with 'objects of use'. Martin Heidegger is one who sees there to be something profoundly important about our interaction with *zeug*, as equipment.[74] We have a specific subjective experience of wielding artefacts and of using them to shape the world.[75] We are, as Thomas Carlyle had it, 'tool-using animals'; the artefacts we create and engage with are part of a distinctive mental architecture. Drawing on 'a neighbouring cultural compartment', we might think the descriptive metaphysician will find something interesting in Henri Bergson's thought, found in *Creative Evolution*, that

> [i]f we could rid ourselves of all pride, if, to define our species, we kept strictly to what the historic and the prehistoric periods show us to be the constant characteristic of man and of intelligence, we should say not *Homo sapiens*, but *Homo faber*.[76]

Though vague, these considerations, and those offered in the previous chapters, encourage the thought that another element of our conceptual framework – alongside our idea of a unified *substance* – is the idea of an *artefact*, an entity that is composed of ontologically independent parts.[77] This proposal will bear refinement.[78] Yet even in its unpolished state it sheds light on the present discussion. Looking at the biological realm mechanistically – in the way in which the brain transplantation narrative invites us to do – we pick out entities with the metaphysical character of *artefacts* and not organisms (*substances*). This explains why Shoemaker's story is so difficult. The idea of brain transplantation is underwritten by a mechanistic logic whereby biological items can be disassembled and reassembled. Organisms cannot be dismantled in this way, and thus the Human Being theorist, who sees personhood to be intimately connected to the idea of a natural substance, cannot hold that Brown survives. Maybe *something* survives. But is it the sort of thing we are normally interested in when posing the questions of personal identity? My answer is that it is not. This reading is fully consonant with Wiggins's texts (indeed, at times, it seems almost to be implicit in his comments about the effects of transplantation),[79] and is, I think, explanatorily generative. Whether he will want to endorse it, however, is another question entirely.

## 6 Final worries and a recommendation for Wiggins

The proposed reading is not without its difficulties. There are two that I would mention here. The first is vague and relates to the internal consistency of my reasoning: Wiggins, I have held, thinks that different entities are caught by different frameworks – but in which framework do we find this 'biological artefact'? The second question is, perhaps, more serious and relates to the overall plausibility of this neo-Aristotelian reading. Pursuing the line advanced above, one may be forced to rule against survival in more commonplace transplantation cases. If, for instance, *heart* transplantation constitutes the disassembly of an organism then on my interpretation we will also say it marks its destruction.

Let us turn to the vague concern about frameworks first and respond to it ... vaguely. It arises because of the claim that different frameworks catch different entities and that the status of the 'biological artefact', as I've called it, is ambiguously placed. 'Biological mechanism' was understood to be part of a scientific world-view (and not our pre-theoretical one). And in some senses, surely, the entity on the surgical slab is a creature that belongs to the mechanist's picture. Simultaneously, Wiggins appears to have the resources to single out this 'living-something-or-other', and to subject it to closer scrutiny. Is he talking about the same thing that appears in the scientific framework or something else? Does the living-something-or-other overlap with the mechanist's subject?

These worries disperse as soon as it is recognized that the divide between science and pre-theory, assumed for argumentative purposes, is less grand, less definitive and perhaps less clear, than it has seemed. The pre-theoretical framework and the scientific one(s) are not discrete,[80] nor separable, though they do sometimes pick out different things (and fail to capture others). So much was hinted at in Chapter 1 where it was pointed out that science is a *human* practice. It is rooted, as all of our endeavours are, in our pre-theoretical framing. This is not to say that these frameworks should be collapsed into each other, but rather to emphasize that real links run between them.

Intriguingly, the comments about 'mechanism' suggest a way of examining the connexion between frameworks. Science is a practice that begins with the objects of our everyday awareness. It aims to explain them – and in doing so, it relies on our pre-theoretical concepts.[81] I have suggested that the concept of the *artefact*, of a human-engineered machine, may have a foundational place in our pre-theoretical world-view. And it is undeniable that the metaphor of the artefact, as something we build, as dissolvable into parts, has been central to scientific developments of the Modern period. Mechanisms are things we engage with in our everyday lives – and from William Harvey's pneumatic conception of the vascular system, to more recent thoughts about genetic 'engineering' – the mechanism has been the prism through which countless biologists have understood the realm of the living.[82]

Nor is it inconsequential that the scientific framework affects changes in the pre-theoretical one. In the quotation above, Wiggins describes a conceptual divergence, where advances in technology encourage us to think of ourselves in new ways. Science is not simply a passive observer. As Hacking points out (alongside post-humanists like Donna Haraway and N. Katherine Hayles),[83] biology, with its sister disciplines, is a creative force; it does not simply explain the world, it remodels it. Doing so, it creates new things for us to navigate, new items for us to pick out. The process is unclear to me – and these thoughts broach work about metaphors and cognitive development that lie beyond the reach of the present inquiry – but it seems plausible that, starting from a certain pre-theoretical view of the world, science might end up 'reshaping' that world-view (which is not to say that it destroys the entities with which it began). Wiggins and the mechanist, though working in different frameworks, can focus on the same object because these frameworks, and the concepts at play, are joined via the metaphor of the mechanism. (That, at least, is the claim I would make if there were no strictures about rigour or evidence in Anglophone philosophy.)

Let us move onto the second question posed above. How does the proposed reading impact upon more 'banal' cases of disassembly like heart transplantation? Much depends on how we are to understand 'disassembly'. Disassembly is not simply the detaching of parts. For that happens routinely and unproblematically (for example, skin or hairs). Nor is disassembly the separation of living body parts. The organism is not disassembled when it loses an eye. Such a loss does not critically undermine the living activity of the organism, even though it impairs it. The human being can happily (or unhappily) suffer minor dental or orthopaedic procedures without any real effect on its distinctive living activity.[84] So 'disassembly' might best be construed as the separation of *vital* organs – where what is 'vital' can be refined, reciprocally, by reference to the physiological-immunological view described in Chapter 6.

However 'vital' is finally cashed out, the heart is clearly among the vital organs. So it seems that heart transplantation will have to count as disassembly. And if Wiggins follows the proposal above he will have to rule *against* the survival of organisms that undergo this apparently 'life-saving' procedure. Moreover – taking his Human Being Theory seriously – the same will be true of the *person*. Despite powerful intuitions to the contrary, the claim will be that persons cannot have their hearts removed and reinserted.[85]

There are two lines of response that Wiggins might take. Both have a degree of textural support in his work. First, he may *accept* the counter-intuitive claim that persons cannot survive heart transplantation. In 'Locke, Butler and the Stream of Consciousness', he almost appears to anticipate this conclusion (and hints too, towards a similar metaphysical analysis of artefacts). He argues there that extensive technological manipulation of the natural substance will cause the person to go out of existence. Importantly, he states that this kind of technological manipulation may be realized in a science-fiction thought experiment, or 'even a kind of practical experiment':[86]

> [It] literally denatures the subject. In place of an animal or organism with a clear principle of individuation one finds an artefact whose identity may be a matter of convention or even caprice. Certainly we do not, at this limit, find a person, if my account of the concept *person* is correct.[87]

We need not wander too far into the realms of science fiction to find examples of this kind of technological interference and Wiggins seems at least open to the possibility that heart transplantation may stand as a limit to personhood. If he thinks that puzzles about artefact identity are a result of their susceptibility to disassembly, and if he thinks that heart transplantation constitutes disassembly, then he might well see the beings that undergo such procedures as demoted to the status of mere artefacts. And since – according to the strictures of **D** – an organism cannot *become* an artefact, the organism (and thus the person) goes out of existence. Of course, we will find it hard to persuade the post-operative patient that she is a different person from the pre-operative one, but, as Wiggins points out in the case of fission, testimony in these cases is contestable.[88]

This is a severe response which many – including Wiggins – will very likely want to resist. It seems clear that a person *can* undergo this kind of heart surgery. Fortunately, there is a more moderate response. It is less counter-intuitive, but it marks a significant departure from Wiggins's official position on personal identity. Persuaded by the argument above, Wiggins may concede that the organism – as a natural *substance* – ceases to exist when it is dismantled; yet disturbed by the harsh rulings on heart transplant survivors, he may ultimately overrule the claims of the Human Being Theory and hold that the person continues. Here, this response joins the critique presented in Chapters 4 and 5. I argued there that the conceptual connection Wiggins traces between the pre-theoretical *person* concept and the concept *human being* (the concept of a natural substance) is weaker than he envisages. *Person*, it was suggested, need not be construed as a *substance* sortal. It may instead be seen to be a *phase* of a kind of biological being (putatively human beings). *Person* might thus encapsulate persistence conditions different from those of *human being*.

The considerations about disassembly put even greater strain on the bond between the two concepts. We feel considerable discomfort at the thought that persons cannot undergo heart transplantation, yet there are strong reasons for denying that human beings – as natural substances – survive them. Furthermore, reflecting on the *Strawsonian argument* and the *argument from interpretation* (presented in Chapter 4), one will wonder whether the claim that the concept *person* must really be tied to the notion of a genuinely unified *substance*. What stops us interpreting, and being interpreted by, entities whose parts do not ontologically depend upon them?[89] The discussions above suggest that we already do.

The thought here is that the conceptual connection, the heartstring of the Human Being Theory, may be severed. Yet Wiggins might be less surprised – if not untroubled – by this thought than one might think. Turning to his more recent work, we find a creeping doubt about the concordance of *person* and *human being*. Recall the dark portents entered at the end of *S&SR*: 'the conception of a human person will diverge further and further from that of a self-moving, animate living being.... The conception will converge more and more closely upon the conception of something like an artefact.'[90] These comments certainly seem to suggest that he is no longer as sure as he once was that the idea of a *person* – a subject of interpretation – is, or must be, the idea of a natural substance.

Moreover, loosing these concepts from one another, Wiggins need not utterly disown his Human Being Theory. While *human being* and *person* are no longer concordant, they might once have been. We may think that conceptual frameworks can *change*, and that, following both biological and cultural evolution, different connections may be made between concepts.[91] We might further think that our understanding of the world, and of ourselves, changes in response to technological advances (which is not to say that we are in thrall to the edicts of science).[92] The possibility of organ transplantation makes us see the world differently; it makes us see ourselves *artefactually*.[93] The Human Being Theory then, which is built around the concordance of *human being* and *person*, may be seen to capture – and capture beautifully – a particular moment in the history of our conceptual framework. But that moment has passed. New connections are being forged. New things are being picked out. Transplantation, grounded in a mechanistic logic, encourages us to connect the *person* concept with the concept of a biological *artefact*, and no longer the concept of an *organism*.[94,95]

## Notes

1  Wiggins 1967: 43.
2  Wiggins 1967: 51.
3  Rory Madden develops this line ('The Persistence of Animate Organisms' (*draft*)). Madden states that the brain carries a sufficient number of capacities of the characteristic activity (15) and as a result holds that we would travel with it (in cases of brain extraction). Conversely, the claim here is that if a biological item survives transplantation it cannot be an organism.
4  Wiggins 1967: 51.
5  See also Wiggins's review of his earlier work, in *S&SR*:

> [Brownson] was the functional inheritor and continuator of all of Brown's vital faculties. *This* was the reason why Brownson counted as the unique inheritor of the title to be Brown, the reason why Brownson was Brown, that very substance. Neither Brown nor Robinson nor Brownson *was* a brain. But the brain being the seat of memory and consciousness was not just any old part of the body among others. It was the essential nucleus of a person (of a human being).
>
> (Wiggins 2001: 207)

6  Wiggins 1967: 52. Note however, Wiggins's suspicions about the supposed symmetry, in *ISTC*. '[I]n the case of a man's brain the two halves of it are not equal in status and that if a surgeon separated them one half would be clever and the other moronic' (1967: 52). This point is reinforced in an interesting footnote in 2001 (208n.21). This is another 'practical' obstacle to the thought experiment which philosophers sideline for the sake of ease – but at the cost of plausibility.
7  Parfit's famous reading of this situation (1971) is that it legislates the separation of the concept of survival from that of identity.
8  Wiggins 1967: 55–58.
9  Wiggins 2001: 208.
10  Wiggins 2001: 209.
11  How do the vital functions 'sit' in the brain? In 1967 this metaphor of the 'seat' is explained in distinctly Aristotelian terms. Indeed, Wiggins sees his response to transplantation to have been anticipated in the *Metaphysics*, and in a significant footnote he imagines Aristotle's response to Shoemaker's story. At *Metaphysics* 1035b14, Aristotle describes certain parts of organisms, 'which are neither prior nor posterior but logically simultaneous with the *psuche* itself, *such as are conceptually indispensible to its existence (*kuria*) and in which the whole formula itself, the essential substance, is immediately present* ... e.g. perhaps the heart or the brain' (see Wiggins 1967: 78). Wiggins applies this, by saying that the brain is not the *psuche* (person) but that it is 'logically simultaneous' with it. Its *logos* (formula) is the *logos* of the *psuche* (form), which is to say that 'its functional mission embraces *everything* [that

is, all the relevant human capacities] which is integral to the *psuche* itself' (1967: 77–78). The brain is the 'seat' of these various capacities because it is casually indispensible to them (in a way that no other organ is). The formula that outlines its function incorporates all those vital functions essential to the *psuche*. For this reason – because its 'formula' is co-extensive with the 'formula' of the form – Wiggins reads the brain as the individuative nucleus of Brown. This interpretation of Aristotle is more than a little controversial (see Ackrill 1997). More will certainly be said by those with a surer grasp of Aristotle's position, but it is interesting, here, to see how the distinction between formula and form translates to the latter distinction in Wiggins's work between 'principle of activity' and 'activity'. The same line is being drawn, and it reveals an interesting ambiguity in Wiggins's account. Does the activity itself need to continue uninterrupted for the individual to survive – or is it enough that the specific determination of the principle of activity (the *logos*) is preserved in some way, with the possibility of it being 're-activated'? The 1967 text does not offer up an immediate answer. Nor is much gleaned from Wiggins's comments in *S&SR* (2001: 207, 226). On an optimistic interpretative note, there seem to be *some* grounds for saying that he *does* hold that the activity must be seamlessly continuous. When considering brain transplantation he writes that the recipient must 'inherit' the epistemic and other capacities 'in the manner in which any ordinary person who has suffered no such adventures is constantly inheriting from himself' (Wiggins 2001: 226). In the normal run of things we 'inherit' these faculties through their continued activity (not 'exercise' necessarily, but 'actualization'). Further, he writes that a human being's persistence requires 'the participation of the same *continued* life' (2001: 207, my emphasis) – though admittedly 'participation' is not wholly transparent here.

12 Williams 1956.
13 Wiggins 1976: 158.
14 Wiggins 1980: 189.
15 Wiggins 1980: 188.
16 E.g. Wiggins 1980: 156 (see also Wiggins 1979).
17 There is a good overview of his concerns in the revised version of his interview with TPM in *Twelve Essays* (2016: fn. 5). Wiggins also suggests that the publication of 'personal identity' put pressure on the 'one parcel stipulation' (2001: 209).
18 Wiggins 2001: 246.
19 Wiggins 2001: 208.
20 Shoemaker 2004a.
21 Wiggins 2004b: 608, fn. 9.
22 Wiggins 2001: 209.
23 Wiggins 2001: 207, fn. 18.
24 Wiggins 2001: 238.
25 Wiggins 2001: 241. And here, perhaps, we may take 'organic independence' to correlate to the kind of ontological independence that organisms were seen to exemplify in Chapter 8.
26 Wiggins 2001: 241.
27 Wiggins 2001: 241.
28 Wiggins 2004a: 605.
29 With respect to this denaturing, Wiggins performs, at the end of *S&SR*, a fascinating shift to a moral register. Even if brain transplantation *were* possible we might think that it shouldn't be:

> If we cannot recognize our own given natures and the natural world as setting any limit at all upon the desires that we contemplate taking seriously; if we will not listen to the anticipations and suspicions of the artefactual conception of human beings that sound in half-forgotten moral denunciations of the impulse to see

people or human beings as things, as tools, as bearers of military numerals, as cannon-fodder, or as fungibles … then what will befall us? Will a new disquiet assail our desires themselves, in a world no less denuded of meaning by our sense of own omnipotence than ravaged by our self-righteous insatiability?… I frame the question and, having framed it, I grave it here.

(2001: 242)

This passage – infused with its neo-Aristotelian morality – merits closer attention. It is a warning – but of what? Of construing others as artefacts? A cursory glance at the feminist literature on the subject of objectification will show that these worries are occasioned by other things than technological advancements (e.g. Langton 2009). (Is the confusion between the natural and the artefactual *necessarily* a bad thing? Much more may be said about this – but it will suffice here to direct attention to Donna Haraway's compelling thoughts about the political power of the notion of a *cyborg*. Disturbing the boundaries between natural and artefactual, the cyborg might allow us to challenge problematic essentializing tendencies (see Haraway 1991) (though, as N. Katherine Hayles (1999) points out, subjects like the cyborg might similarly be *re-inscribed* with problematic notions of, for example, Cartesian dualism).).

30  Wiggins 2004a: 604–605.
31  Wiggins 2012: 20.
32  Wiggins 2012: 20. The revised (2016) version of the paper supports some of these complaints. He focuses on the 'co-adaptation', achieved over time, between the nervous system and the limbs. Can a brain be inserted, without issue, into a body with whom it has no history? Maybe not – but if not, Wiggins will have to explain why this 'co-adaptation' is more important with brains than for other organs.
33  This is a lack of clarity that Lowe also identifies (Lowe 2003: 819).
34  Wiggins 2004a: 605. Madden (2011) focuses on a different kind of shift that occurs in Shoemaker's scenario – a *referential shift*. The reference of 'I' shifts slowly and undetectably from Brown to Brownson – or, as he puts it, from the 'Old Animal' to the 'New Animal' (Madden 2011: 296). I am not sure what to say about this, but perhaps – given the arguments below – that shift might better be construed as one between an organism and an artefact.
35  Le Doeuff 1989/2000, James 2000, la Caze 2002.
36  Their approach can be contrasted with Kathleen Wilkes's (1988) who dismisses science fiction thought experiments because they are unfavourable to reflective equilibrium. The relevance of this thought to the present discussion becomes clear in her comments about fissioning:

> [I]n a world where we split like amoebae, everything else is going to be so unimaginably different that we do not know what concepts will remain "fixed", part of the background; we have not filled out the relevant details of this "possible world", except that we know it cannot be much like ours. But if we cannot know that, then we cannot assess, or derive conclusions from, the thought experiment.
>
> (1988: 12)

37  La Caze 2002: 2. La Caze (and, for example, James 2000: 30–32) also points out that these thought experiments – and this naïve approach to them – are perhaps most common in Analytic philosophy.
38  Shoemaker 1963: 23–24.
39  See Wiggins 2001, Chapter 7 §7, 1996: 246, 1980: 188–9, 1967: 53.
40  James 2000.
41  James 2000: 33.
42  James 2000: 32ff. Nor, of course, is it irrelevant that the story's protagonists are all male.
43  James 2000: 33.

44 These thoughts bear on the personal identity discussion about 'thinking parts'. Madden points out that the claim that we have *human form* is threatened by the commonplace thought that our conscious perspective could, in theory, be had by beings smaller and less humanoid than ourselves (*forthcoming*: 2–3). As 'phantom limbs' demonstrate, the possession of body parts is not necessary for us to *experience* possessing them. Thus the conscious experience of a human being might be had by a thinking part – a brain, say. We have no way of telling which one we are. Madden ultimately rejects the presupposition that 'the local activity in parts significantly smaller than the whole humanoid is sufficient for the presence of a conscious perspective' (7, 23). His conclusion may be supported by the arguments above, which emphasize the extent to which our conscious experience is an experience of an *embodied being*. Amputees may feel things in now lost limbs (what exactly? an itch? a pain? a thrill of excitement? And which limbs? Philosophers tend not to elaborate) – but there is a big step from this to the claim that a thinking part might think the things it thinks *in* a body when it is no longer *part* of that body. Relatedly, the problem of 'remnant persons' – animals pared down to, for example, cerebrums in vats – gains no purchase. Whatever such things might dream, their thoughts will likely be nothing like our own conscious experience (see Wiggins 1996: 246, cf. Madden *forthcoming*: 31).

45 E.g. Shoemaker 2004a: 574.

46 Shoemaker 1963: 23.

47 Wiggins 2001: 92.

48 Note, too, that Brown exemplifies what the social scientist Karl Deutsch sees to be the central features of the classical concept of a mechanism. A mechanism behaves 'in an exactly identical fashion no matter how often [its] parts [are] disassembled and put together again' (Deutsch 1951: 233–234). Madden describes a tendency among philosophers to visualize living organisms along the lines of 'wooden block "anatomical toys" … a collection of interlocking wooden blocks, any one of which may be freely removed and returned' (*forthcoming*: 32). This is the picture I think can be helpfully articulated by reference to mechanism.

49 Hacking 2007.

50 Hacking 2007: 80.

51 Hacking 2007: 102–103.

52 Dupré 2008/2012: 71.

53 This interpretation can be helpfully brought out by a comparison with E.J. Lowe's analysis of a dismantled watch, found in 'On the Identity of Artifacts':

> When Jones's watch goes to the watchmaker for repair and is taken to pieces by him, it doesn't *cease to exist*, does it? Someone entering the workshop and seeing the pieces laid out carefully on the watchmaker's bench would quite properly be told 'That is Jones's watch'; and if such a person were, say, to stamp on these delicate bits of machinery, he would clearly be guilty of *destroying* Jones's watch, i.e. terminating its existence.
>
> (Lowe 1983: 222)

Likewise, if someone was to go into the operating theatre and stamp on the relevant bits, one might also say it was at that point that they killed the patient.

54 The difference between *quantitative* and *qualitative* difference is set out, in relation to anti-reductionism, by Allen (1975: 106).

55 Looking through the theatre window is it plausible to say that patient has *ceased* to exist? Perhaps one is committed to the thought that biological life occurs only when the relevant parts are united? On this reading the unified whole might be seen to exert a relevant causal power over its parts, and one might still resist reductionism. This interpretation leads one down a hazardous path, however; the thesis that existence may be, as Lowe puts it, 'intermittent', or 'interrupted' (Lowe 1983: 222), is a deeply controversial one. It stands in marked contrast to the view, set out in Locke's *Essay*,

that 'one thing cannot have two beginnings of existence' (Locke 1690/1975: II. xxvii.1) – a compelling thought, and one which finds support, most notably, in Wiggins's work ('a thing starts existing only once' (Wiggins 2001: 92)).

56 Wiggins 1980: 167, 2012: 13–14.
57 Here, I think, Wiggins is susceptible to a line of criticism from post-humanist theorists, like N. Katherine Hayles. Hayles argues (1999) that the notion of a unified being, which underpins our moral discourse, is the invention of a particular form of Liberal Humanism (stemming, significantly, from Locke). The notion of a social contract, for example, is fully reliant upon, and reinforces, the thought of agents as unified and discrete (see also Ferner (*draft*)). Is the genuine unity, which – it has been argued – is a central aspect of the pre-theoretical *substance* concept, really a specific sociocultural construct? This is a question, which those wishing to pursue the line of argument advanced above must seriously contend with.
58 Of course, there are presumably some machinic parts that are particular to certain artefacts (e.g. watch batteries) – but not in nearly so fine-grained a way.
59 Pasnau 2004: 42–43. NB This is a more extreme, Thomistic interpretation of Aristotle, where artefacts are non-substances. But the point still stands.
60 One result of the present work is to (hopefully) encourage wariness of this and comparable narratives. When 'neo-Aristotelians' like van Inwagen and Olson describe cases of frozen cats and draw analogies between human bodies and lumps of clay, they need to work out fully the suppressed commitments these stories contain (see, for example, van Inwagen 1990: 146, Olson 2007: 220).
61 Wiggins 2001: 229.
62 Wiggins 2001: 211.
63 Wiggins 2016.
64 This is from the revised form of 'Identity, Individuation, and Substance' (2012) in *Twelve Essays* (2016).
65 Wiggins 2012: 1.
66 Wiggins, unfortunately, does not describe the metaphysical character of concrete universals in any real depth. But he seems to suggest that the notion of a multiply-instantiated object is a part (though a less prominent part) of our conceptual framework (1980: 166–167). They appear to be entities that can be elucidated within his descriptivist framework.
67 The telling title of this section is: 'One Last Variant – and the Philosophical Moral of the Same. Finally, *Human Persons as Artefacts?*' (Wiggins 2001: 236 (my emphasis)).
68 Wiggins 2001: 241 (my emphasis).
69 See, for example, Simons 1987: 197 and, indeed, Wiggins 2001: 10, 87, 91, 2004a: 605.
70 Gainotti 2012.
71 For a fair overview of the topic, and the various positions held within it, I recommend the forum on Krist Vaesen's 'The Cognitive Bases of Human Tool Use', which appeared in *Behavioral and Brain Sciences* (Vacscn 2012).
72 Orban and Rizzolatti 2012.
73 Vaesen 2012. Of course, humans are not the only animals that use tools (though Vaesen argues that, in this regard, humans have a special aptitude).
74 Heidegger 1927/1962:

> In our dealings we come across equipment for writing, sewing, working, transportation, measurement. The kind of Being which equipment possesses must be exhibited. The clue for doing this lies in our first defining what makes an item of equipment – namely its equipmentality.
>
> (I.3 (97))

I have Charlotte Knowles to thank for making this aspect of Heidegger clear(er) to me.

75  See, for example, Botvinick and Cohen 1998.

76  Bergson 1911/1998: 139.

77  What is the relationship between having a principle of functioning and having onto-logically independent parts? Wiggins states that one correlate of having a principle of functioning is that, when picking out artefacts, one need make no claims about their specific constitution (2001: 87). Having a nomologically shallow essence, *clock* picks out items that work in radically different ways and are made of radically different things. There is thus no intimate connection between our understanding of the *parts* of a clock and our understanding of what a clock is.

78  One will wonder at what point the 'living artefact' comes into existence. Is it only *after* the technological interference? Or is the artefact's birth the same as the organism's? This question is just as pressing in the case of the concrete universal produced/revealed by fission, and bears too on Wiggins's brief discussion of *objet trouvé* (2001: 134 fn. 38). At what point does Duchamp's *Fountain* – metaphysically distinct from the urinal – come into existence? The onus is on Wiggins to explain these cases. Perhaps he has strong grounds for stating that the surgeon has 'created a living something-or-other' (and not unmasked one) but he has yet to explain them. He may be helped here by Madden's discussion of 'creation and exposure' (*forthcoming*: 32ff.).

79  Wiggins 2001: 241, 1976: 154.

80  There is an issue here, which has been suppressed, but which is signaled by the bracket 's': how many scientific frameworks are there? This depends on whether or not one buys into the project of scientific unity. Perhaps, in the end, we *will* be able to explain all the physical phenomena we find around us by referring to the tiny interact-ing parts. Perhaps the reductionist's thesis will be rescinded in the light of discoveries about the causal powers of higher-level beings? Either way, this is for the scientists and the science-led metaphysicians to decide; Wiggins's descriptivism is untouched by these concerns and so for present purposes both positions are assumed to be viable.

81  See (again) Wiggins 2012: 24, fn. 29.

82  There are innumerable analyses of the effects of the mechanism metaphor but one of the best, and the one which informs the present study, is Haraway's *Crystals, Fabrics, and Fields: Metaphors of Organicism in Twentieth-Century Developmental Biology* (1976). See also Dupré 1993.

83  I am thinking, particularly, of Hayles's *How We Became Posthuman* (1999), and Har-away's 'Cyborg Manifesto' (1991).

84  Wiggins 2001: 240–241.

85  Still, we can have our spleens and appendices, one (but not both) kidneys removed (and any number of teeth, hairs and eyeballs ...).

86  Wiggins 1976: 154.

87  Wiggins 1976: 154. The term 'denaturing' is interesting – and nowhere replicated in his work. The idea of 'denaturing', of one individual losing its distinctive mode of being and gaining another, is inimical to Wiggins's sortal theory **D** (there being no change in the category of substance). This is another reason why the proposed reading of artefacts (whereby substances do not *become* them, but *coincide* with them) is an improvement.

88  Do we see the appearance here of something like Olson's 'thinking animal puzzle' (2007: 29–39)? The patient who emerges from the operating room after a successful heart transplant has clear memories of entering the theatre – does this mean that prior to the operation there were two thinking things in the same place at the same time? The person (*a natural substance*, on Wiggins's view) and some other one? Olson identifies two worries with this kind of 'cohabitation' (2007: 35–36): the *overcrowd-ing problem* – whereby the number of thinkers in the world is greater than we intuit-ively think – and the *epistemic problem* – where one cannot tell which entity one is. There is not the space to go into these in depth, but some brief thoughts may be entered.

(i) The worry depends on whether the biological artefact comes into existence after the transplant (this line is suggested by Wiggins's comments in 2004a, where he thinks the surgeon *creates* a living something-or-other). Other comments suggest he seems open to the possibility that the artefact dates back to the genesis of the organism (this line is prompted by his analogous discussion (1980) of the *concrete universal* 'Brown' – which dates from Brown's birth, not his fissioning). If one takes the first line there will be no problem. If the latter, then the burden is on Wiggins to answer the worry (to explain how a concrete universal and a substance may cohabit).

(ii) One might offer a deflationary response on Wiggins's behalf. The *overcrowding problem* is not a problem for realists of his ilk. The *epistemic problem* is a problem, but one that should be embraced as accurately capturing the unease we actually feel when faced, for example, with the prospect of extensive radiotherapy, gene therapy, drug therapy, transplantation, etc. We genuinely worry that we may not survive such procedures, that the thing that has suffered these intrusions will no longer be 'me'.

89 Wiggins may point to the necessary open-endedness of our pre-theoretical *person* concept. As discussed in Chapter 4 he denies that we can list the psychological and physical features that are relevant to Davidsonian interpretation. He holds that the list is necessarily 'open-ended' and that the sense of 'person' cannot be stipulated. It must, in some way, be attached to a natural kind, with a nomologically profound *principle of activity*. If artefacts are items with *principles of functioning* – nominally stipulated – then they cannot be persons (understood by reference to the notion of a subject of interpretation) – or so the argument will go. Three brief points in response: First, Wiggins's association of *aposiopesis* and a notion of a subject of interpretation (discussed in Chapter 4) is cursory at best. Second, as Wiggins himself points out (2001: 242) the artefactual view of persons is open-ended anyway (though worryingly so). Third, as discussed Wiggins considers (though only implicitly) the possibility that our concept of *person* may attach to the concept of an *artefact* – so he cannot himself be fully persuaded by this response.

90 Wiggins 2001: 241. It is notable that Wiggins writes 'conception' here and not 'concept'. This is more than a little confusing. As discussed, 'conception' relates to a particular construal of a concept. Whatever the conception, it still picks out the same concept, which picks out the same thing in nature. Is this what Wiggins is describing? Not according to the analysis presented, where artefacts are metaphysically distinct from natural substances. To see persons artefactually is to associate them with a different metaphysical kind.

91 This is one way of interpreting Wiggins's own comments that 'as human beings have come to the point where their powers of reason and analogy make it possible for some of them to transcend mere species loyalty, the sense of *person* has been very slightly modified' (Wiggins 1976: 152).

92 This seems to be suggested by Wiggins's comments at the end of *S&SR* (quoted above).

93 The thought is offered again in his discussion with Shoemaker (2004a: 605).

94 How does this suggestion relate to the discussion of *conceptual invariance*? Not only is there conceptual variation over time, but this analysis suggests there is conceptual variation over cultures. Are there politically problems with holding this? These are thoughts that must be addressed, in detail, at another time.

95 At the risk of repeating myself – but to militate against the greater risk of being misunderstood – let me restate my advice for Wiggins. He should say, I think, that organic substances, understood in the neo-Aristotelian sense advanced above, *cannot* undergo heart transplantation, but that *we* – you and I – most surely can. The neo-Aristotelian substances are beings that we once picked out, with ease, but thanks to

the creative and reconceiving powers of science we have begun to latch onto entities with a different metaphysical character to the kind that Aristotle took as his subject. We see biological artefacts. This is not a result of philosophical factionalism but a consequence of technological advances and their effects, which cause (or are causing) a shift in our conceptual framework. The real connection that once existed between the *person* concept and the concept of an organic *substance*, has been disturbed.

# Bibliography

Ackrill, J.L. (1997) 'Aristotle's Definitions of *Psuche*', in J.L. Ackrill *Essays on Plato and Aristotle* (Oxford: Oxford University Press).

Allen, G. (1975) *Life Science in the Twentieth Century* (New York: Wiley and Sons).

Aristotle. (1994) *Metaphysics* (Books Z and H), D. Bostock (ed.) (Oxford: Clarendon Press).

Bergson, H. (1911/1998) *Creative Evolution*, A. Mitchell (trans.) (New York: Dover Publications).

Botvinick, M. and Cohen, J. (1998) 'Rubber Hands "Feel" Touch That Eyes See', *Nature* 39(1): 756.

Deutsch, K.W. (1951) 'Mechanism, Organism, and Society: Some Models in Natural and Social Science', *Philosophy of Science* 18(3): 230–252.

Dupré, J. (1993) *The Disorder of Things: Metaphysical Foundations of the Disunity of Science* (Cambridge, MA: Harvard University Press).

Dupré, J. (2008/2012) 'The Constituents of Life 1: Species, Microbes, and Genes', in J. Dupré *Processes of Life* (Oxford: Oxford University Press).

Ferner, A. (*draft*) 'Posthumanism's Troubled Historicity'.

Gainotti, G. (2012) 'Comment on Vaesen', in the forum on 'The Cognitive Bases of Human Tool Use', *Behavioral and Brain Sciences* 35(4): 203–262.

Hacking, I. (2007) 'Our Neo-Cartesian Bodies in Parts', *Critical Inquiry* 34(1): 78–105.

Haraway, D. (1976) *Crystals, Fabrics, and Fields: Metaphors of Organicism in Twentieth-Century Developmental Biology* (New York: Yale University Press).

Haraway, D. (1991) 'A Cyborg Manifesto: Science, Technology, and Socialist-Feminism in the Late Twentieth Century', in D. Haraway *Simians, Cyborgs and Women: The Reinvention of Nature* (London: Routledge).

Hayles, N.K. (1999) *How We Became Posthuman* (Chicago: University of Chicago Press).

Heidegger, M. (1927/1962) *Being and Time*, J. Macquarrie and E. Robinson (trans.) (Oxford: Basil Blackwell).

James, S. (2000) 'Feminism in Philosophy of Mind: The Question of Personal Identity', in M. Fricker and J. Hornsby (eds) *The Cambridge Companion to Feminism in Philosophy* (Cambridge: Cambridge University Press).

la Caze, M. (2002) *The Analytic Imaginary* (New York: Cornell University Press).

Langton, R. (2009) *Sexual Solipsism: Philosophical Essays on Pornography and Objectification* (Oxford: Oxford University Press).

le Doeuff, M. (1989/2000) *The Philosophical Imaginary* (Stanford: Stanford University Press).

Locke, J. (1690/1975) *An Essay Concerning Human Understanding*, P.H. Nidditch (ed.) (Oxford: Oxford University Press).

Lowe, E.J. (1983) 'On the Identity of Artifacts', *Journal of Philosophy* 80(4): 220–232.

Lowe, E.J. (2003) 'Review of *Sameness and Substance Renewed*', *Mind*, New Series, 112 (October): 448.

Madden, R. (*draft*) 'The Persistence of Animate Organisms'.

Madden, R. (2011) 'Intention and the Self', *Proceedings of the Aristotelian Society* 111(3): 325–351.

Madden, R. (*forthcoming*) 'Thinking Parts', in S. Blatti and P. Snowdon (eds) *Essays on Animalism* (Oxford: Oxford University Press).

Olson, E. (1997) *The Human Animal: Personal Identity Without Psychology* (Oxford: Oxford University Press).

Olson, E. (2007) *What Are We? A Study in Personal Ontology* (Oxford: Oxford University Press).

Orban, G. and Rizzolatti, G. (2012) 'Comment on Vaesen', in the forum on 'The Cognitive Bases of Human Tool Use', *Behavioral and Brain Sciences* 35(4): 203–262.

Parfit, D. (1971) 'Personal Identity', *Philosophical Review* 80: 3–27.

Parfit, D. (1984) *Reasons and Persons* (Oxford: Oxford University Press).

Pasnau, R. (2004) 'Form, Substance, and Mechanism', *The Philosophical Review* 113(1): 31–88.

Shoemaker, S. (1963) *Self-Knowledge and Self-Identity* (Ithaca: Cornell University Press).

Shoemaker, S. (2004a) 'Brown-Brownson Revisited', *The Monist* 87(4): 573–593.

Shoemaker, S. (2004b) 'Reply to Wiggins', *The Monist* 87(4): 610–613.

Simons, P. (1987) *Parts: A Study in Ontology* (Oxford: Oxford University Press).

Vaesen, K. (2012) 'The Cognitive Bases of Human Tool Use', *Behavioral and Brain Sciences* 35(4): 203–262.

van Inwagen, P. (1990) *Material Beings* (Ithaca: Cornell University Press).

Wiggins, D. (1967) *Identity and Spatio-Temporal Continuity* (Oxford: Blackwell).

Wiggins, D. (1976) 'Locke, Butler and the Stream of Consciousness: And Men as a Natural Kind', *Philosophy* 51: 131–158.

Wiggins, D. (1979) 'Mereological Essentialism: Asymmetrical Dependence and the Nature of Continuants', in E. Sosa (ed.) *Essays on the Philosophy of Roderick Chisholm* (Amsterdam: Grazer Philosophische).

Wiggins, D. (1980) *Sameness and Substance* (Cambridge, MA: Harvard University Press).

Wiggins, D. (1996) 'Replies', in S. Lovibond and S. Williams (eds) *Essays for David Wiggins: Identity, Truth and Value* (Blackwell Publishing).

Wiggins, D. (2001) *Sameness and Substance Renewed* (Cambridge: Cambridge University Press).

Wiggins, D. (2004a) 'Reply to Shoemaker', *The Monist* 87(4): 594–609.

Wiggins, D. (2004b) 'Reply to Shoemaker's Reply', *The Monist* 87(4): 614–615.

Wiggins, D. (2012) 'Identity, Individuation and Substance', *European Journal of Philosophy* 20(1): 1–25.

Wiggins, D. (2016) *Twelve Essays* (Oxford: Oxford University Press).

Wilkes, K. (1988) *Real People: Personal Identity Without Thought Experiments* (Oxford: Oxford University Press).

Williams, B. (1956) 'Personal Identity and Individuation', *Proceedings of the Aristotelian Society* 57.

# Conclusion

It is over 50 years since David Wiggins first began his exploration into individuation, identity and substance. His thoughts have changed and ranged, but where sometimes his work is hard to follow it is always easy to track, characterized by his distinctive style and insight. His dealings with these matters enthral and enrich those who attend to them – and the intention here has been to offer both an introduction and a development of them.

Among other things, I have tried to show exactly *why* Wiggins's work is so difficult to read. His investigations intersect and interrelate more closely than most – his account of personal identity is carefully constructed alongside, and in conjunction with, his thoughts about the logic of identity, our individuative procedures and his sober brand of conceptualist-realism. In this – his *systematicity* – he resists the dominant impulse in English-language philosophy for Russellian piecemeal analysis; as a result, his work is hard to grasp ... but hard to grasp for the right reasons. The fine-spun links he draws between these different issues represent the real and important connections he sees to hold between our pre-theoretical concepts. That is, the complexity in his texts mirrors the complexity he divines in our minds.

It was the work of Chapters 1, 2 and 3 to demonstrate the close connections Wiggins finds between our thoughts about *identity, individuation* and *substance*. He was seen there to hold that our everyday ability to navigate the world is underwritten by these concepts, and it was argued that his critics and commentators fail when they do not appreciate how his analyses of these notions relate to one another. I have tried to correct these commentaries.

Another central aim of Chapter 3 was to assess his view, and the various interpretations of it, of the distinction between *natural things* and *artefacts*. Misinterpretations were identified and a reading was offered whereby he was seen to hold that artefacts – stipulatively defined – are substances, but never paradigms of that category. I suggested that Wiggins has the resources to analyse better the ontological profile of artefacts, so that they may be differentiated from natural substances.

This proposal was temporarily shelved, and in Chapter 4 I turned my attention to Wiggins's Human Being Theory. In line with his sortalism, he holds that, in order to answer the puzzles of the personal identity debate, we must turn to the sortals under which we ourselves fall and interrogate the *principle of activity* that they encapsulate. The ingenuity of his Human Being Theory is to argue for a conceptual consilience of *person* (which, despite long-standing discussion, remains elusive) and *human being* (a notion with which we seem to be much more familiar). Having argued for this concordance, Wiggins claims that both concepts have the same underlying *principle of activity*, to which we should defer when ruling on persistence.

The conceptual connection is grounded in three lines of argument: the *Strawsonian argument*, the *semantic argument* and the *argument from interpretation*. Following a genealogical critique in Chapter 5, I rejected the semantic argument. As a result the other two arguments appeared much weaker. The conceptual consilience that Wiggins finds between *person* and *human being* is less stable than he envisages.

Another concern that arose was that the *human being principle* might be more difficult to explicate than one might suppose. In Chapter 6 – following comments by Wiggins – it was suggested that one should turn to *biology* to investigate the *living activity* in greater depth. But biological individuation was found to be a controversial affair. I argued that Wiggins and the animalists must choose between a variety of biological models if they want to set the parameters for an investigation into the principle of activity for the human being. Certain models – the immunological view being one – were seen to offer some support, but a secondary worry then appeared, arising from the 'counter-intuitive' picture of heterogenous organisms.

In Chapter 7, these thoughts about principles of activity led onto broader questions about metaphysical reductionism. The debate between the emergentists and the reductionists was laid out, and I advanced an emergentist reading of Wiggins but found it wanting. An alternative response to the threat of reductionism was offered in Chapter 8, where the focus was turned to the metaphysical character of the *natural substances*. It was argued that Wiggins might benefit from endorsing a neo-Aristotelian view of organisms, with Kantian notes. Organisms exemplify the category of substance because they are *genuine unities*. They are genuine unities because, when we latch on to them, we cannot help but see them as *ontologically prior* to their parts.

Chapters 6, 7 and 8 show what Wiggins might glean (as he always anticipated) from the insights of philosophers of biology and – by presenting an alternative form of anti-reductionism – how he might contribute to their discussions as well. Beneath the fields of biology and metaphysics are deep, living roots – roots that are no less strong for being hidden – and one thought I have intended to promote is that inter-disciplinary discussion will be mutually beneficial (a thought further encouraged by the connection, described in the Appendix, between the neo-Lockean account of personal identity and seventeenth-century mechanism).

At the same time, I see the discussions above to encourage scepticism about inter-disciplinarity projects – specifically those conducted between metaphysicians and biologists. Aside from the practical obstacles, there are, perhaps, metaphysical ones as well. The biologists study certain kinds of entities, while the metaphysicians (at least some of them) study others. This is not to dissuade discussion across the spheres but to urge caution and care when doing so.

It was in Chapter 9 that these threads were drawn together. The analysis of the metaphysical character of organisms offered in Chapter 8 was seen to bear on the distinction, discussed in Chapter 3, between natural things and artefactual things. The tentative suggestion entered there – that one may draw a principled line between *substances* and *artefacts* – was revisited and prompted a re-examination of Shoemaker's notorious brain transplantation story. Fleshing out the distinction between the ontological profile of artefacts and substances provided an explanation of why Shoemaker's narrative is so unsettling. The notion of brain transplantation shifts our metaphysical focus from a biological *substance* onto a biological *artefact*.

This thought is connected to the conclusion I reached in Chapter 5: that the conceptual consilience of *person* and *human being* is not as strong or resilient as Wiggins supposes. The concordance of concepts that stands as the core of his Human Being Theory was taken to be undermined. Yet rather than disavow his theory *tout court*, it was suggested that one should instead say that it marks a significant moment in the history of our conceptual framework. The methods that underpin Wiggins's analysis are in no way challenged by the arguments above – but being sensitive to our everyday thoughts, and the fluctuations of our conceptual framework, they now license a different conclusion to the one encapsulated by the Human Being Theory.

At the close, I find I can echo a sentiment voiced by Wiggins at the end of his first published paper, 'The Individuation of Things and Places' (1963). It is an expression of a dissatisfaction, verging on a hope – one that will resonate with all those who take as their subject rich and elusive works. It is a promise of the changes one could make and the limitations of what one has written – a conditional too often disappointed, but one which Wiggins has thoughtfully fulfilled, and one which I will leave unfinished in the happy expectation of comments to come:

If I had time to rewrite my contribution ...

# Appendix
## A history of the brain transplantation story

This appendix is intended as a supplement to the work in the final chapter, to show that there is, perhaps, a non-accidental connection between Shoemaker's neo-Lockeanism and the mechanistic notion of a disassemblable human body. It presents a speculative history of the brain transplantation story, focusing on how shifts in seventeenth-century biology may have influenced Locke's development of his *person* concept.

There is a textual trail that connects Shoemaker to the mechanist debates of the seventeenth century. Shoemaker explicitly states that his story of Brown and Robinson is a modern retelling of Locke's 'Prince and Cobbler' narrative.[1] Locke, in turn, states that his story presents a resolution to a problem caused by the Christian doctrine of the resurrection.[2] Locke's endorsement of a causal form of biological mechanism – his corpuscularianism – will be described. Then it will be argued that the problem with resurrection is a problem for Locke precisely *because of* his corpuscularianism (and his side-lining of immaterial substances). Then it will be claimed that Locke's corpuscularianism intersects with the parallel, mechanist thesis that biological parts are conceivable in isolation from whole bodies. Lastly, it will be suggested that the brain transplantation story – like Locke's story of the thinking finger – is mechanistic in this latter sense, since it is by thinking of situations where consciousness survives in separated body parts that the *person* concept demonstrates its strengths.[3]

## 1 Locke's mechanistic/corpuscularian picture

Insofar as *materialism* is the thesis that only material things exist, Locke is not a materialist. He believes in the existence both of God and of souls.[4] Does this mean he thinks these immaterial things play a role in natural processes? If he does, what kind of role is it? Is God positioned as a primary mover who sets up the natural system and leaves it to run according to geometrical necessity and intelligibility, or does Locke hold that the natural realm will collapse without God's constant miracle-working influence?

The first thing to mention is Locke's moderate scepticism in response to such questions. There is, he thinks, a limit to human understanding, and some phenomena in the natural realm lie beyond it. A case in point is the connection between primary and secondary qualities; why some primary qualities cause in us specific ideas – of colours, of pains, etc. – is, Locke thinks, beyond our ken. Perhaps the connection is determined by mechanical necessity, but perhaps it is forged by God to meet some other demands.[5] This being said, there is strong evidence to suggest that, when it comes to living beings, Locke thinks they are best understood in *corpuscularian* terms, as the products of the interaction of textured, indivisible particles, or 'corpuscules'. He does not seem to think that God has any immediate, active role in the continual process and development of nature. Thus, in the *Essay* at least, he advances the mechanical view advanced by Robert Boyle.[6]

In the late seventeenth century, Boyle's brand of corpuscularianism was one of the dominant physical theories in England.[7] In 'Excellency and Grounds of the Mechanical or Corpuscular Hypothesis' (1674) and *Origin of Form and Qualities*, he defends the view that the natures of material objects (including biological ones) arise solely from 'the size, shape, motion (or want of it), texture and the resulting qualities of the small particles of matter'[8] that make them up. For a corpuscularian like Boyle the world is composed of these basic units of matter – '*minima naturalia*' – arranged in certain ways,[9] which, in virtue of their textural and motive differences, impart the qualities and properties found in material objects. Thus, Jonathan Walmsley writes:

> Boyle's view, as stated in the published works that we know Locke to have read, was that God had created the world as a uniform matter divided into variously shaped and textured particles. These pieces of matter interacted with each other according to mechanical laws. The only part of nature that Boyle held not to be mechanical was Man and his rational soul.[10]

Locke, like Boyle, affirms the existence of God and rational souls. Yet, on one reading at least, God is invoked only as a prime mover[11] and the soul adds nothing to the explanation of organic life, but is a special addition in the case of humans. The biological picture that Locke endorses is thus, on its face, similar to the kind of mechanistic one endorsed by the Cartesians (and Gassendians),[12] with their *bête-machine*.[13] (The only notable point on which both Locke and Boyle differ from Descartes, is the question of plenism: they posit indivisible atoms and void, rather than the plenum of matter, which Descartes identifies with extension.[14])

Boyle's mechanistic materialist view of the natural world thus stands in opposition to the views of the Aristotelian scholastics and the vitalists.[15] The Aristotelians – who dominated Oxford at the time[16] – advocated various forms of hylomorphism and endorsed a similar kind of organicist picture to the one described in Chapter 8.[17] The vitalists – like Henry More, Francis Glisson and Joan Baptista van Helmont[18] – posited vital spirits to explain biological phenomena (van Helmont, for example, understood biological life by reference to 'Archeus', the active spirit, or 'seminal' principle, that guided the actions of organisms).[19]

Walmsley suggests that in Locke's early work there are signs he may have subscribed to a form of vitalism – but he notes, too, that by the time of writing of the *Essay*, Locke demonstrably endorses the Boylean brand of corpuscularianism.[20] In the *Essay* – the focus of this discussion – he eschews talk of substantial forms, and teleology, and vital spirits, and conceives the biological realm as the product of the interaction of minute, moving particles.[21] There, at least, the organism is conceived of as 'nothing more than the sum of its parts'. It is seen as a mechanism, exemplifying the causal thesis above, as something that derives its causal powers from its constituents and the relations that hold among them.

## 2 The problem of resurrection

It is in the second edition of the *Essay* that Locke presents the story of the prince and the cobbler. He does so in order to demonstrate how his position can combat a puzzle raised by the Christian doctrine of the resurrection.[22] What is this puzzle and why does it affect corpuscularians but not their Scholastic, Vitalist or Cartesian contemporaries? The worry is raised in detail in Boyle's 'Some Physico-Theological Considerations About the Possibility of the Resurrection'. Boyle writes:

> When a man is once really dead, divers of the parts of his body will, according to the course of nature, resolve themselves into multitudes of steams that wander to and fro in the air; and the remaining parts, that are either liquid or soft, undergo so great a corruption and change, that it is not possible so many scattered parts should be again brought together, and reunited after the same manner, wherein they existed in a human body whilst it was yet alive. And much more impossible it is to effect this reunion, if the body has been, as it often happens, devoured by wild beasts or fishes; since in this case … they are quite transmuted as being informed by the new form of the beast or fish that devoured them and of which they now make a substantial part.[23]

He continues:

> And yet far more impossible will this reintegration be, if we put the case that the dead man was devoured by cannibals; for then, the same flesh belonging successively to two different persons, it is impossible that both should have it restored to them at once, or that any footsteps should remain of the relation it had to the first possessor.[24]

It is significant that these worries about cannibals will not arise for the Aristotelian-scholastics or for the vitalists; both posit some theoretical principle, *over and above* the material parts, by which humans can be individuated. The various parts of the body may well resolve into steams and disperse, but the Aristotelians will refer to the form, or *psuche*, of living things,[25] and the vitalists, like van Helmont, will refer to some vital spirit, to explain how an individual may be traced through the resurrection. The material mixing of parts will not result in the mixing or dissolution of *Archei* or seminal forces or substantial forms.

The resurrection of the body becomes a problem once the emphasis, in individuation, is put exclusively on the material parts of which the body is made. And the difficulty of tracing human bodies through resurrection *without* endorsing these kinds of non-reductionist models is clear from Boyle's own unsatisfying responses to the problem. His first recourse is to turn to the durability of certain body parts. While the body is in perpetual flux, he writes, there are some parts that are of a 'stable and lasting texture', such as the bones.[26] Since bones can suffer fire and other assaults they may, Boyle suggests, provide a material basis for individuation.[27] At other points he suggests that corpuscles can *retain* their original nature under various disguises, in the way that gold particles may be dissolved into solutions, and resolved at a later point into gold.[28] As K.J.S. Forstrom notes in her thorough study of the debate, all of these responses were seen to be largely unpersuasive. And Locke, while attending closely to Boyle's discussion, pointedly takes a different route in accommodating the resurrection doctrine.

Locke diverges from Boyle in changing the focus of the question. What is at issue, he claims, is the resurrection of the 'dead', not necessarily their bodies. This move appears in his correspondence with Bishop Stillingfleet,[29] and relies on the crucial distinction that Locke draws between *man* (human) and *person*.[30] In Chapter 27 of Book II, Locke argues that 'man' and 'person', while sometimes used synonymously, do not refer to the same thing. On seeing a creature of our 'own Shape and Make', though with 'no more reason all its Life, than a *Cat* or a *Parrot*', we would undoubtedly call it a 'Man' (human), but (he says) not a 'Person'. And finding a cat or a parrot, discoursing, reasoning and philosophizing, we would call them persons, though not men.[31] And – in line with the now familiar sortalism – both 'Man' and 'Person' have, according to Locke, distinct principles of individuation:

> Locke distinguishes between the identity of the self as man (or human being) and the identity of the self as *person*. The identity of the self as man consists in the identity of the same bodily organism; the identity of the self as person, by contrast, is constituted by the consciousness of our thoughts and actions.
>
> (Essay II, xxvii.16)[32]

This distinction allows Locke to bypass the questions about bodily continuity through the resurrection. He interprets the account in the Paul's epistles as concerning the *person* – because it is the person and not the living body which is the object of reward or punishment. He writes (quoting from Corinthians):

> the Apostle tells us, that at the Great Day, when every one shall *receive according to his doings, the secrets of all Hearts shall be laid open*. The Sentence shall be justified by the consciousness all Persons shall have, that they *themselves* in what Bodies soever they appear … are the *same*, that committed those Actions, and deserve Punishment for them.[33]

It is this move that is illustrated, explicitly, by the story of the prince and the cobbler, in section 15 of Chapter 27:

> And thus we may be able without any difficulty to conceive, the same Person at the Resurrection, though in a Body not exactly in make or parts the same which he had here, the same consciousness going along with the Soul that inhabits it. But yet the Soul alone in the change of Bodies, would scarce to any one, but to him that makes the Soul the *Man*, be enough to make the same *Man*. For should the Soul of a Prince, carrying with it the consciousness of the Prince's past Life, enter and inform the Body of a Cobler as soon as deserted by his own Soul, every one sees, he would be the same Person with the Prince, accountable only for the Prince's Actions: But who would say it was the same Man?[34]

There may be vestiges of scholasticism here – in the description of the Soul *informing* (that is, being the *form* of) the body – but more importantly, the story seems overtly Cartesian in tone. The emphasis is put on an immaterial soul, which, it is implied, is the thinking substance, *res cogitans*. This, certainly, is one method for circumnavigating the problem raised by the resurrection. As Udo Thiel puts it:

> For most of those thinkers who believe that the soul is an immaterial substance, there is no real problem of personal identity at all. They would argue that personal identity consists in the identity of a mental substance or soul and that the identity of a mental substance is a direct consequence of its immaterial nature; it is because of its immateriality that the mind is not subject to change and remains the same through time.[35]

Crucially, however, Locke *denies* that personal identity consists in the identity of the soul. He specifically argues against the elision of the person and a mental substance. The story of the Prince and the Cobbler is *not* an illustration of a Cartesian response to the resurrection. Alongside the distinction between man and person, Locke distinguishes between person and soul. This is the focus of his discussion in Chapter 27 of the 'rational Man' who believed himself the reincarnation of Socrates, but could remember none of Socrates' actions:

> Let him also suppose it to be the same Soul, that was in *Nestor* or *Thersites* … which it may have been, as well as it is now, the Soul of any other Man: But he, now having no consciousness of any of the Actions either of *Nestor* or *Thersites*, does, or can he conceive himself the same Person with either of them? Can he be concerned in either of their Actions? Attribute them to himself, or think them his own more than the Actions of any other Man, that ever existed?[36]

While Locke believes in the existence of the soul, he also argues that its principle of continuity differs from the principle of continuity for persons. There is room for interpretation here, depending on whether or not Lockean 'persons' are understood as genuine substances or modes of other substances (or virtual substances).[37] But on one prominent reading, Locke takes consciousness to be a property of another substance while remaining determinately neutral as to whether it will be a material or immaterial one.[38] Thus, though he holds it to be the 'more probable Opinion' that consciousness is 'annexed' to the soul,[39] he writes pointedly that it involves 'no contradiction' to think that *matter* might have been made by God fitly disposed to think.[40] Despite its description of soul transference, the Prince and the Cobbler story is not a Cartesian one. It is because of the transfer of consciousness that the prince wakes to find himself in the cobbler's quarters.[41]

Locke's response to resurrection is novel. Because of his advocacy of Boylean corpuscularianism he cannot explain the identity of the pre- and post-resurrection individual by reference to a common vital spark or substantial form. And because he denies the soul is identical to the person, and side lines the immaterial from his account, he will not explain it by reference to a common mental substance. His method – which has become one of the most influential in Anglophone philosophy[42] – is to focus on the *consciousness* of persons.[43] Resurrection involves the resurrection of persons, and the identity of persons is constituted by continued consciousness, rather than organic continuity, or continuity of an immaterial substance.

## 3 A connection

Locke's position is a subtle one. He needs to highlight the distinctiveness of the person's persistence conditions in contrast to those of the man/human being and the soul. Dialectically, then, he is moved to present a situation where the animal does not continue, and where the persistence of the immaterial soul can also be called into question. This is exactly the kind of situation he describes in section 17 of Chapter 27:

> [E]very one finds, that whilst comprehended under that consciousness, the little Finger is as much a part of it *self*, as what is most so. Upon separation of this little Finger, should this consciousness go along with the little Finger, and leave the rest of the Body, 'tis evident the little Finger would be the *Person*, the *same Person*; and *self* then would have nothing to do with the rest of the Body. As in this case it is the consciousness that goes along with the Substance, when one part is separated from another, which makes the same *Person*, and constitutes this inseparable *self*.[44]

The story of the *thinking finger* demonstrates the strengths of the *person* concept, and by linking personhood to a *material part of the body* it distinguishes continuity of consciousness from organic continuity, and from the continuity of an immaterial substance. Significantly, Locke's 'thought experiment' also demonstrates the interplay between a causal form of mechanism – corpuscularianism – and the mechanist thesis (described above), which relates to ontological dependence. For the causal mechanists, the biological whole exerts no novel causal influence over its parts. Holding this, they are not *necessarily* committed to the view that body parts are *ontologically independent* of the animal; still, it is a view with which they may well be sympathetic. If one thinks that biological phenomena result from uni-directional causal chains reaching up from constituent corpuscles one can readily accept that animal parts may exist when separated from animal wholes. Indeed, taking the causal thesis to its extreme, one finds the kind of reductionist scenario described by Empedocles (discussed in Chapter 8), that peculiar stage in cosmic history where heads and trunks and limbs are produced independently and roll around to produce bizarre, chimerical, 'scrambled' animals.[45] The point is simply that two forms of mechanism intersect and both seem at play in Locke's story of the *thinking finger*. His description of that body part contrasts notably with Aristotle's description of the same:

[T]he finger cannot exist apart from a living animal.[46]

Locke's *thinking finger* is a somewhat forgotten item of the philosophical imaginary – but it points to an important way in which issues of personal identity are bound up with particular notions of biology. The Lockean concept of a person, developed to cope with puzzles encouraged by causal mechanism, demonstrates its strengths in a situation where the body is construed *mechanistically*. Such a situation involves the transmission of consciousness in a material body part, like a thinking finger – or, equally, a brain.[47]

## Notes

1 Shoemaker 2004: 573, 1963: 21.
2 Locke 1690/1975: II.xxvii.15 (and Uzgalis 2012).
3 It is important, once again, to stress the spirit in which this is offered. It is not intended as a work of historical scholarship so much as a speculative attempt to articulate what seem to me to be real connections between certain ideas in different, related philosophical contexts. With respect to any historical claims, I am relying heavily on the work of experts like Udo Thiel, Jonathan Walmsley, Michael Ayers and K. Joanna Forstrom – the aim is to focus on the dots they point to, and try to join them to the ones I have found in the contemporary Anglophone personal identity debate.
4 E.g. Locke 1690/1975: II.xxvii.2.
5 Ayers gives a helpful overview of these and related issues (1991: Chapters 11 and 12).
6 This is the reading, which I take to be relatively uncontroversial, proposed by (among others) Ayers (1991), Brun (2007), Thiel (1998).
7 Kochiras 2009.
8 Boyle 1674: VIII, 104.
9 Forstrom 2010: 104.
10 Walmsley 2000: 375.
11 Forstrom 2010: 104, Walmsley 2000: 377.
12 Forstrom 2010: 102.

13 E.g. Martin and Barresi 2000: 14. This is a rather simplistic reading of both Boyle and Descartes. It is, as Cottingham notes, not clear that Descartes himself ever really endorsed the thesis that animals are, metaphysically speaking, nothing more than clocks. Cottingham 1978 (cf. Ferner 2008).

14 Kochiras 2009.

15 It is, however, worth noting the blurriness of the boundaries between each of these groupings. At the start of Chapter 7, the overlapping of the methodological, epistemic and metaphysical was emphasized, and this is clear in the different kinds of 'corpuscularianism'. Consider, for example, Daniel Sennert's avowedly Aristotelian corpuscularianism (Walmsley 2000: 369).

16 Forstrom 2010: 6.

17 Martin and Barresi 2000: 14.

18 Walmsley 2000: 370.

19 Boyle too spoke of a 'seminal' principle – but Walmsley states this was purely mechanical:

> Boyle did not suppose the seminal principles to be anything other than textured matter in special relationships with their surroundings by the laws of the nature and the circumstances in which they are placed. There is no evidence that he believed these agents to act in anything other than a mechanical way.
>
> (Walmsley 2000: 378)

20 Walmsley 2000. Walmsley argues that in his early 'Essay on Disease', Locke advanced something like Helmont's vitalism (2000: 381).

21 Martin and Barresi 2006: 123.

22 The standard rending of the resurrection, which Locke and his contemporaries focused on, was that given in the epistles of St. Paul (Forstrom 2010: 2).

23 Boyle 1675: 198.

24 Boyle 1675: 198.

25 Martin and Barresi 2000: 14.

26 See Forstrom 2010: 110ff. One interesting response to Boyle has been brought to my attention by Justin Smith – the pressure cooker was developed by Denis Papin not long after the *Essay* was published; it was notable for being able to reduce bones to edible sludge (Leibniz, indeed, wrote a satirical letter to his employer, complaining about this invention, on behalf of the court dogs).

27 Boyle offers a second response along similar lines, which refers to the original matter of the foetus (Forstrom 2010: 110–111).

28 Forstrom 2010: 111.

29 Thiel 2011: 134–135.

30 Forstrom 2010: 112ff.

31 Locke 1690/1975: II.xxvii.8.

32 Thiel 1998: 61.

33 Locke 1690/1975: II.xxvii.26.

34 Locke 1690/1975: II.xxvii.15.

35 Thiel 1998: 62. See also Martin and Barresi 2000: 13–14.

36 Locke 1690/1975: II.xxvii.14.

37 See Martin and Barresi 2000: 24–25 for an outline of the controversy.

38 Thiel 2011: 144. NB The use of 'substance' here is slightly orthogonal to the Aristotelian use I have advanced in the rest of this book, since on the reading I have presented, neither Wiggins nor Aristotle would recognize 'immaterial' substances.

39 Locke 1690/1975: II.xxvii.25.

40 Locke 1690/1975: IV.iii.6. For the widespread influence of this suggestion, see Yolton 1983 and Thiel 1998.

41 Thiel 1998: 61.

42 There will be multifarious reasons for this, but not least among them is the Analytic canonization of Locke by Russell and Ryle (see Akehurst 2010 – in which he quotes Russell's comment to Ryle: 'By God ... I believe you are right. No one ever had Common Sense before John Locke – and no-one but Englishmen have ever had it since' (1)).

43 Uzgalis 2012.

44 Locke 1690/1975: II.xxvii.17.

45 Furth 1987: 44.

46 Aristotle *Metaphysics* 1036b. Also 'the finger is defined by the whole body. For a finger is a particular kind of part of a man' (*Metaphysics* 1035b).

47 The aim here has been to highlight certain important disagreements between Aristotle's picture and Locke's. Yet there are important agreements too; the neo-Lockean account of personal identity is not so inimical to an Aristotelian framework as some suppose (for more on this see Whiting 2008).

# Bibliography

Akehurst, T. (2010) *The Cultural Politics of Analytic Philosophy: Britishness and the Spectre of Europe* (Continuum).

Aristotle. (1994) *Metaphysics* (Books Z and H), D. Bostock (ed.) (Oxford: Clarendon Press).

Ayers, M. (1991) *Locke*, vol. 2 (London: Routledge).

Brun, C. (2007) 'Lockean Mechanism and the Principle of Identity', *Graduate Faculty Philosophy Journal* 28(2).

Boyle, R. (1674) 'The Excellency and Grounds of the Mechanical Hypothesis', in R. Boyle *The Excellency of Theology* (London: Henry Herringman).

Boyle, R. (1675) 'Some Physico-Theological Considerations About the Possibility of the Resurrection', in M.A. Stewart (ed.) *Selected Philosophical Papers of Robert Boyle* (Manchester University Press, 1979).

Cottingham, J. (1978) '"A Brute to the Brutes?": Descartes' Treatment of Animals', *Philosophy* 53(206): 551–559.

Ferner, A. (2008) 'Review of Cottingham's *Cartesian Reflections*', *Philosophy* 85(4): 580–584.

Forstrom, K.J.S. (2010) *John Locke and Personal Identity: Immortality and Bodily Resurrection in Seventeenth-Century Philosophy* (London: Continuum).

Furth, M. (1987) 'Aristotle's Biological Universe: An Overview', in A. Gotthelf and J.G. Lennox (eds) *Philosophical Issues in Aristotle's Biology* (Cambridge: Cambridge University Press).

Kochiras, H. (2009) 'Locke's Philosophy of Science', in E.N. Zalta (ed.) *The Stanford Encyclopedia of Philosophy* available online at http://plato.stanford.edu/archives/fall2009/entries/locke-philosophy-science/

Locke, J. (1690/1975) *An Essay Concerning Human Understanding*, P.H. Nidditch (ed.) (Oxford: Oxford University Press).

Martin, R. and Barresi, J. (2000) *Naturalization of the Soul: Self and Personal Identity in the Eighteenth Century* (London: Routledge).

Martin, R. and Barresi, J. (2006) *The Rise and Fall of Soul and Self: An Intellectual History of Personal Identity* (New York: Columbia University Press).

Shoemaker, S. (1963) *Self-Knowledge and Self-Identity* (Ithaca: Cornell University Press).

Shoemaker, S. (2004) 'Brown-Brownson Revisited', *The Monist* 87(4): 573–593.

Thiel, U. (1998) 'Locke and Eighteenth-Century Materialist Conceptions of Personal Identity', *The Locke Newsletter* 29: 59–83.

Thiel, U. (2011) *The Early Modern Subject: Self-Consciousness and Personal Identity from Descartes to Hume* (Oxford: Oxford University Press).

Uzgalis, W. (2012) 'John Locke', in E.N. Zalta (ed.) *The Stanford Encyclopedia of Philosophy*, available online at http://plato.stanford.edu/archives/fall2012/entries/locke/.

Walmsley, J. (2000) '"Morbus" – Locke's Early Essay on Disease', *Early Science and Medicine* 5(4): 366–393.

Whiting, J. (2008) 'The Lockeanism of Aristotle', *Antiquorum Philosophia*.

Yolton, J. (1983) *Thinking Matter: Materialism in Eighteenth-Century Britain* (Minneapolis: University of Minnesota Press).

# Index

# Taylor & Francis eBooks

## Helping you to choose the right eBooks for your Library

Add Routledge titles to your library's digital collection today. Taylor and Francis ebooks contains over 50,000 titles in the Humanities, Social Sciences, Behavioural Sciences, Built Environment and Law.

**Choose from a range of subject packages or create your own!**

**Benefits for you**

» Free MARC records
» COUNTER-compliant usage statistics
» Flexible purchase and pricing options
» All titles DRM-free.

**Benefits for your user**

» Off-site, anytime access via Athens or referring URL
» Print or copy pages or chapters
» Full content search
» Bookmark, highlight and annotate text
» Access to thousands of pages of quality research at the click of a button.

REQUEST YOUR **FREE** INSTITUTIONAL TRIAL TODAY

**Free Trials Available**
We offer free trials to qualifying academic, corporate and government customers.

## eCollections – Choose from over 30 subject eCollections, including:

| | |
|---|---|
| Archaeology | Language Learning |
| Architecture | Law |
| Asian Studies | Literature |
| Business & Management | Media & Communication |
| Classical Studies | Middle East Studies |
| Construction | Music |
| Creative & Media Arts | Philosophy |
| Criminology & Criminal Justice | Planning |
| Economics | Politics |
| Education | Psychology & Mental Health |
| Energy | Religion |
| Engineering | Security |
| English Language & Linguistics | Social Work |
| Environment & Sustainability | Sociology |
| Geography | Sport |
| Health Studies | Theatre & Performance |
| History | Tourism, Hospitality & Events |

For more information, pricing enquiries or to order a free trial, please contact your local sales team: www.tandfebooks.com/page/sales

 Routledge
Taylor & Francis Group

The home of Routledge books

www.tandfebooks.com